U0483790

符号中国 SIGNS OF CHINA

二十四节气

TWENTY-FOUR SOLAR TERMS

"符号中国"编写组 ◎ 编著

中央民族大学出版社
China Minzu University Press

图书在版编目(CIP)数据

二十四节气:汉文、英文 /"符号中国"编写组编著. —北京:
中央民族大学出版社,2024.3
（符号中国）
ISBN 978-7-5660-2303-2

Ⅰ.①二… Ⅱ.①符… Ⅲ.①二十四节气—介绍—中国—汉、英 Ⅳ.①P462

中国国家版本馆CIP数据核字（2024）第016838号

符号中国：二十四节气 TWENTY-FOUR SOLAR TERMS

编　　著	"符号中国"编写组
策划编辑	沙　平
责任编辑	杨爱新
英文指导	李瑞清
英文编辑	邱　械
美术编辑	曹　娜　郑亚超　洪　涛
出版发行	中央民族大学出版社
	北京市海淀区中关村南大街27号　　邮编：100081
	电话：（010）68472815（发行部）　传真：（010）68933757（发行部）
	（010）68932218（总编室）　　　　（010）68932447（办公室）
经 销 者	全国各地新华书店
印 刷 厂	北京兴星伟业印刷有限公司
开　　本	787 mm×1092 mm　1/16　印张：11.875
字　　数	154千字
版　　次	2024年3月第1版　2024年3月第1次印刷
书　　号	ISBN 978-7-5660-2303-2
定　　价	58.00元

版权所有　侵权必究

"符号中国"丛书编委会

唐兰东　巴哈提　杨国华　孟靖朝　赵秀琴

本册编写者

王　佳

前言 Preface

　　中国是世界上最早发明历法的国家之一，二十四节气是中国历法的重要组成部分。中国有着灿烂辉煌的农业文化，二十四节气则是农耕文明的重要体现。二十四节气是古代劳动人民对天文、气候、农业生产等各方面

China is the earliest country to apply the calendric system in the world of which the Twenty-four Solar Terms constitute a significant part. Ancient China was mainly founded on the basis of agriculture and therefore created a brilliant agricultural culture. And the Twenty-four Solar Terms

经验的总结，是指导农事活动的重要依据，至今仍与人们的日常生活密不可分。

本书从节气的由来、相关的气象物候、农事活动、民间风俗、健康等方面，对二十四节气做了系统的介绍和解读，帮助读者了解中国独特的历法以及节气对生活、农事生产等方面的影响。

precisely reflect this distinct culture. The Twenty-four Solar Terms are conclusions on the study and experience regarding to astronomy, climate and agricultural production, it has been a major reference to conducting farm work and so far has a close relationship with people's daily lives.

This book gives a systematic introduction and explanation of the Twenty-four Solar Terms in terms of its origin and relative knowledge of weather and climate, phonological phenomena, agricultural activities, folk customs and health preservation to help the audience gain further understanding of this unique calendric system and its great impact on several aspects of Chinese people's life and agricultural production.

目 录 Contents

二十四节气概述
General Introduction to the
Twenty-four Solar Terms.................................. 001

二十四节气的来历
Origin of the Twenty-four Solar Terms............ 002

二十四节气的类别
Classification of the
Twenty-four Solar Terms.................................. 011

二十四节气与中国人的生活
Twenty-four Solar Terms
and Chinese People's Daily Life 015

二十四节气
Twenty-four Solar Terms 019

立春
Beginning of Spring (*Lichun*) 020

雨水
Rain Water (*Yushui*) 029

惊蛰
Awaking of Insects (*Jingzhe*) 036

春分
Spring Equinox (*Chunfen*) 043

清明
Pure Brightness (*Qingming*) 049

谷雨
Grain Rain (*Guyu*) 060

立夏
Beginning of Summer (*Lixia*) 069

小满
Grain Buds (*Xiaoman*) 077

芒种
Grain in Ear (*Mangzhong*) 083

夏至
Summer Solstice (*Xiazhi*) 091

小暑
Minor Heat (*Xiaoshu*) 096

大暑
Major Heat (*Dashu*) 102

立秋
Beginning of Autumn (*Liqiu*) 108

处暑
End of Heat (*Chushu*) 116

白露
White Dew (*Bailu*) 121

秋分
Autumn Equinox (*Qiufen*) 128

寒露
Cold Dew (*Hanlu*) 135

霜降
Frost's Descent (*Shuangjiang*) 141

立冬
Beginning of Winter (*Lidong*) 146

小雪
Minor Snow (*Xiaoxue*) 152

大雪
Major Snow (*Daxue*) 157

冬至
Winter Solstice (*Dongzhi*) 162

小寒
Minor Cold (*Xiaohan*) 168

大寒
Major Cold (*Dahan*) 172

二十四节气概述

General Introduction to the Twenty-four Solar Terms

  古代劳动人民按照太阳的运行规律安排作息，日出而作，日落而息。随着古代天文学的发展，人们首先将黄道附近的星象按方位划分为四象二十八宿，奠定了确立二十四节气的基础。

  二十四节气反映了季节的变化，指导着农事活动，影响着百姓的日常生活，是中华民族世代相传的宝贵文化遗产。

Ancient Chinese people arranged their schedule according to the Sun's movement as they started one day's farm work with the sunrise and rested with the sunset. Accompanying the development of ancient astronomy, people first defined the Four Symbols and Twenty-eight Mansions around the ecliptic, known as the foundation of the Twenty-four Solar Terms.

  The Twenty-four Solar Terms reflect the seasonal changes, provide important guidance for agricultural activities and influence the daily life of ordinary people, respected as the precious heritage passed down from generation to generation by Chinese people.

> 二十四节气的来历

古代劳动人民根据太阳运行到达太阳黄经的位置以及地面气候演变次序来确定二十四节气。地球公转一周为360°，二十四节气是将这360°平均分为24份。太阳从黄经（指太阳经度或天球经度，是在黄道坐标系统中用来确定天体在天球上位置的坐标值）0°起，沿黄经每运行15°所经历的时日即为一个节气。

西周时期（前1046—前771）农业、手工业的长足进步，促进了早期天文学的发展，人们首先确定了春、夏、秋、冬四季。春秋中期，人们使用圭表观测日影，即在地上垂直树立一个标杆，称为"表"，在旁边南北方向水平放置一把尺子，称为"圭"。当太阳照射到表

> Origin of the Twenty-four Solar Terms

Ancient laboring people defined the Twenty-four Solar Terms according to the position relative to the ecliptic longitude, and the change of seasons. The 360° revolution of the Earth is equally divided into 24 parts which are known as the Twenty-four Solar Terms. From the 0° ecliptic longitude (referring to the solar or celestial longitude, known as the coordinate figure in the ecliptic coordinate system to determine the position of celestial bodies on the celestial sphere), every 15° the Sun's movement is regarded as one solar term.

In the Western Zhou Dynasty (1046 B.C.-771 B.C.), the great progress of agriculture and handicraft industry promoted the development of early astronomy. Ancient Chinese people first determined the four seasons as spring,

上的时候，影子就会落在圭上，人们通过影子移动的规律、影子的长短来观测季节变化，并利用水均衡滴漏的原理，发明了漏壶来确定时刻。人们将日影最长和最短的日子

• 日晷
Sundial

summer, autumn and winter. During the middle period of the Spring and Autumn Period (770 B.C.-476 B.C.), people began to employ the Earth sundial to observe seasonal variation, consisting of a stick vertically placed on the ground as the *Biao* (gnomon) and a ruler horizontally located beside the gnomon in both south and north directions, called as the *Gui* (sundial). When the Sun shines on the gnomon, the shadow will be presented on the sundial and through its regular movement pattern and length, people can measure the Sun's shadow. In addition, the clepsydra was invented to indicate the time based on the principle of water balance. At that time, people defined the days with the longest and shortest Sun's shadow as the Winter Solstice and Summer Solstice. When it comes to the Zhou Dynasty (1046 B.C.-256 B.C.), the Beginnings of Spring, Summer, Autumn and Winter were further determined and in the Qin and Han dynasties (221 B.C.-220 A.D.), the other 16 solar terms were ascertained and widely applied to the agricultural production.

The Twenty-four Solar Terms were officially recorded in *Book of Huainanzi: Patterns of Heaven* during the early years of the Western Han Dynasty (179 B.C.-

定为冬至和夏至，将昼夜时间相等的两天定为春分和秋分。周朝时，又用这个方法确立了立春、立夏、立秋和立冬四个节气。秦汉时期，确立了另外十六个节气，并应用于农业生产。

二十四节气正式的文字记载，出现在成书于西汉初年（前179—前121）的《淮南子·天文训》中。公元前104年，落下闳、邓平等人制定的《太初历》将二十四节气作为历法收录，明确了二十四节气的天文位置。人们根据太阳在黄道上位置的变化规律，总结出黄河中下游地区，在一个回归年（从地球上看，太阳围绕天球的黄道一周的时间，即太阳两次经过春分点所经历的时间，又称"太阳年"）中，天文、气象、物候和农事活动等方面的规律和特征，并依此来指导农事。

二十四节气的制定发源于黄河中下游地区。黄河流域是中国古代农业文明的发源地之一。黄河中下游地区位于北纬30°至40°，一年四季气候分明，阳光水分充足，而且地势平坦，土地肥沃，非常适宜耕作。从二十四节气形成到完备的时期里，黄河中下游地区一直是古

121 B.C.). In 104 B.C., Luoxia Hong and Deng Ping with other professionals defined and recorded the Twenty-four Solar Terms as the calendar in *Taichu Calendar*, and clarified the astronomical position of the Twenty-four Solar Terms. According to the rule of the Sun's position change along the ecliptic, Chinese people concluded the regulation and characteristics of the astronomy, weather, phenology and agricultural activities for the regions of the middle and lower reaches of the Yellow River within one tropical year (the length of time that the Sun takes to move around the ecliptic of the celestial sphere, as seen from the Earth and also known as a solar year). Therefore, it is also the calendar to guide agricultural activities.

The creation of the Twenty-four Solar Terms originates from the regions of the middle and lower reaches of the Yellow River, known as the place of origin of ancient agricultural culture in China. These regions are located in the area between the 30° and 40° northern latitude and have distinct four seasons all year round, with adequate sunshine and rainfall, flat landscape and fertile soil, which are all wonderful conditions for cultivation. With the development

● 山西乾坤湾

黄河流域是中华民族的发祥地之一，乾坤湾地处黄河中游山西黄土高原腹地，是远古祖先生息繁衍的中心区域之一。

Qiankun Bay in Shanxi Province

The Chinese nation originates from the Yellow River basin and the Qiankun Bay is located in the hinterland of the Loess Plateau along the midstream of the Yellow River, known as one of the central areas for ancient Chinese ancestors to live and multiply.

代中国政治、经济和文化的中心。二十四节气基本概括了自然现象的变化规律，记载了自然中的物候现象（指动植物生命活动的季节性变化以及一年中特定时间出现的气象、天文现象），反映了黄河中下游地区的气候条件和农业特点，对农业发展起到了积极的促进作用。二十节气指导性强，通俗易懂，因此在中国各地都因地制宜得到了广泛的应用。

节气在公历中的日期基本固定，分布在一年12个月里，上半年在每月的6日、21日前后，下半年在8日、23日前后，前后只差1~2天。每月有两个节气，一个在前半月的月初，俗称"节气"，一个在后半月的月中，俗称"中气"。在农历中，平年每月也是两个节气，一个是节气，一个是中气。若遇到农历闰年的闰月就只有一个节气，没有中气。节气和中气交替出现，各历时15天，现在人们已经把节气和中气统称为"节气"。

节气有：立春、惊蛰、清明、立夏、芒种、小暑、立秋、白露、寒露、立冬、大雪和小寒；中气

of the Twenty-four Solar Terms, these regions have been treated as the political, economic and cultural centre of China. The Twenty-four Solar Terms generally summarise the changing law of natural phenomena, record the phonological phenomena (referring to the seasonal changes of animals and plants, as well as the meteorological and astronomical phenomena of certain periods in a year), and reflect the climatic conditions and agricultural characteristics of the regions of the middle and lower reaches of the Yellow River, playing a positive role in promoting the agricultural development. Since this calendar is valuable in guiding agricultural activities, the Twenty-four Solar Terms were popularised in China at that time, and widely applied to local conditions.

The dates of the Solar Terms are fixed in the Gregorian calendar, distributed in twelve months of a year, known as the 6th and 21st of each month in the first half year and the 8th and 23rd of each month in the second half year, with one to two days' difference. There are two Solar Terms in a month: the one at the beginning of the first half month is called as the first terms (*Jieqi*) and the other one in the middle of the latter

有：雨水、春分、谷雨、小满、夏至、大暑、处暑、秋分、霜降、小雪、冬至和大寒。

half month is known as the middle terms (*Zhongqi*). It is the same with the lunar calendar, and there are two solar terms

二十四节气歌

春雨惊春清谷天，夏满芒夏暑相连，
秋处露秋寒霜降，冬雪雪冬小大寒。
上半年是六廿一，下半年是八廿三。
每月两节日期定，最多相差一两天。

Ballad of Twenty-four Solar Terms

The first half of this ballad is composed with a string of individual characters (each took from the corresponding Chinese term to together represent the Twenty-four Solar Terms in chronological order for people to remember); and the second half indicates their approximate dates in the Gregorian calendar.

Ballad of Twenty-four Solar Terms

Spring begins, it rains, waking insects; after Spring Equinox, come Pure Brightness and Grain Rain.

Beginning of Summer, Grain Buds, Grain in Ear, Summer Solstice, Minor Heat and Major Heat are sequential.

After Beginning of Autumn, End of Heat and White Dew, it is Autumn Equinox, and then Cold Dew and Frost's Descent.

Beginning of Winter, Minor Snow, Major Snow, Winter Solstice, Minor Cold and Major Cold come one by one.

The end of the first half year is the 21st day of the sixth lunar month.

The beginning of the second half year is the 23rd day of the eighth lunar month.

Each month has two solar terms.

The exact date is around within one or two days.

in one month on average, also known as the first terms and middle terms. With regards to the leap month in lunar leap year, there is only one Solar Term without the middle term. In general, first terms and middle terms appear respectively with a duration of 15 days. At present, people unify the first and middle terms as the Solar Terms.

The Twenty-four Solar Terms include the first terms and middle terms which respectively appear at the beginning of the month (around the 6th or 8th) and at the end of the month (21st and 23rd). The first terms are Beginning of Spring (*Lichun*), Awakening of Insects (*Jingzhe*), Pure Brightness (*Qingming*), Beginning of Summer (*Lixia*), Grain in Ear (*Mangzhong*), Minor Heat (*Xiaoshu*), Beginning of Autumn (*Liqiu*), White Dew (*Bailu*), Cold Dew (*Hanlu*), Beginning of Winter (*Lidong*), Major Snow (*Daxue*), Minor Cold (*Xiaohan*). And the middle terms are Rain Water (*Yushui*), Spring Equinox (*Chunfen*), Grain Rain (*Guyu*), Grain Buds (*Xiaoman*), Summer Solstice (*Xiazhi*), Major Heat (*Dashu*), End of Heat (*Chushu*), Autumn Equinox (*Qiufen*), Frost's Descent (*Shuangjiang*), Minor Snow (*Xiaoxue*), Winter Solstice (*Dongzhi*), Major Cold (*Dahan*).

- 《菊花螃蟹图》齐白石（近代）
Chinese Painting: *Chrysanthemums and Crabs*, by Qi Baishi (Modern Times)

公历、农历

我国现行的日历是公历,全称"太阳历",是当今世界上大多数国家、地区和民族通用的历法。公历是以地球绕太阳公转的周期来计算的。地球绕太阳一周所用的时间为一个回归年,是公历的基本周期,一个回归年是365日5小时48分46秒,而公历年是365日,为了补足差数,历法规定,每4年中有一年另加1日,为366天,称"闰年"。每年的4、6、9、11月是小月,30天;2月为28天,闰年2月为29天;其余月份是大月,31天。

农历是我国传统历法之一,属于阴阳合历,目前与公历并行使用。农历以朔望(朔是指月球处于太阳与地球之间,从地球上看不见月亮;望是指地球位于太阳与月球之间,从地球上看月球呈圆形)的周期来定月,用置闰的办法使年平均长度接近太阳回归年。由于这种历法安排了二十四节气以指导农业生产,故称农历。农历早在夏朝就已制定,又称"夏历"。民间也有称为"阴历"的。农历把日、月合朔的日期作为月首,即初一,把朔望月(月亮圆缺的一个周期)作为历月的平均时间。农历以12个月作为一年,共354日或355日,与回归年相差11日左右。每隔3年多出33天,为了消除多余的日数,每隔3年加1个月,就是农历的闰月。农历平年354天或355天,闰年为383天或384天。

Gregorian Calendar and Lunar Calendar

At present, the calendar applied in China is the Gregorian calendar, known as the Solar Calendar, the universal calendar employed by most countries, regions and nationalities in the world. The Gregorian calendar is calculated based on the Earth's period of revolution around the Sun. The length of time for the Earth to move around the Sun is a tropical year and the basic cycle of the Gregorian calendar of 365 days 5 hours 48 minutes and 46 seconds. However, one Gregorian year is 365 days. It is regulated that for one in every four years, one day will be added to the leap year of 366 days to deal with the time difference. April, June, September and November are defined as the Lesser Months of 30 days; ordinarily, February has 28 days and the one in the leap year has 29 days; the rest months are called the Greater Months consisting of 31 days.

The lunar calendar is one of the traditional calendars in China, belonging to the lunisolar calendar, and is applied to the Gregorian calendar at the same time. The lunar calendar employs the period of *Shuowang* (syzygy) to define the month (*Shuo* refers to when the Moon is between the Sun and Earth, and the Moon cannot be seen from the Earth. *Wang* indicates when the Earth is between the Sun and Moon, the Moon is presented in a round viewed from the Earth), and the appearance of the leap month is based on the purpose of making the length of a year close to that of a tropical year. Since the Twenty-four Solar Terms in this calendar are the guidelines of agricultural activities, it is known as the lunar calendar. The lunar calendar was introduced and applied as early as the Xia Dynasty (approx. 2070 B.C.-1600 B.C.), also called the Summer Calendar or Lunar Calendar, prevailing among ordinary Chinese people. In the lunar calendar, the syzygies of the Sun and Moon are regarded as the beginning of a month, known as the first day of the lunar month, and the synodic month (one phase of the Moon) is the average time duration of one lunar calendar month. A lunar year consists of 12 months of 354 or 355 days in total, with approximately 11 days' difference compared with a tropical year. For every three years, there are 33 excess days and in order to deal with them, one month will be added for every three years, treated as the lunar leap month. Therefore, one lunar common year consists of 354 or 355 days and one lunar leap year has 383 or 384 days.

> 二十四节气的类别

二十四节气按照不同的性质，可以分为四个类别：季节类、气温类、降水类、物候类。

季节类：立春、春分、立夏、夏至、立秋、秋分、立冬、冬至。这八个节气是对季节变换的反映。立春表示春季到来，万物将要进

> Classification of the Twenty-four Solar Terms

The Twenty-four Solar Terms can be classified into four types according to natures, including season, temperature, precipitation and phenology.

According to season: Beginning of Spring, Spring Equinox, Beginning of Summer, Summer Solstice, Beginning of Autumn, Autumn Equinox, Beginning of

● 江苏扬州春景
Spring Scenery of Yangzhou City of Jiangsu Province

入萌芽、生长的季节；春分代表春季已经过去一半，昼夜时间均等，在此之后白昼的时间将会变长；立夏代表夏季到来，天气将会变得炎热，植物繁茂，动物开始进入生长繁殖期；夏至代表天气正式变得炎热，春季已经一去不复返；立秋代表秋季来临，天气将会慢慢变得凉爽，农作物进入收获的季节；秋分表示秋季过去一半，在此之后白昼时间将会越来越短；立冬代表冬季到来，天气将会变得寒冷，动物蛰伏过冬，而植物也开始衰败；冬至表示寒冷的冬季正式来临。

气温类：小暑、大暑、处暑、小寒、大寒。这五个节气是对气温变化的反映。小暑表示天气开始炎热；大暑表示天气已经到达全年气温最高点；处暑表示炎热的天气即将结束；小寒表示寒冷天气来临；大寒表示天气已经到达全年气温的谷底。

降水类：雨水、谷雨、白露、寒露、霜降、小雪、大雪。这七个节气反映的是全年降水的变化。雨水代表干燥少雨的冬季已经结束，全年降雨季即将到来；谷雨代表降

Winter and Winter Solstice. These eight Solar Terms reflect the seasonal changes. Beginning of Spring indicates the arrival of spring and all plants begin to sprout and grow. Spring Equinox means that half of spring has passed away and the length of day and night is the same, after which the length of daytime increases. Beginning of Summer marks the arrival of summer and during this period, the weather becomes hotter, plants flourish and animals step into the phase of growth and reproduction. Summer Solstice represents that it is burning hot and the spring ends and Beginning of Autumn means the autumn comes. The weather during this period turns increasingly cool and crops enter the harvest season. Autumn Equinox refers to that half of the autumn has passed by and after that, the daytime becomes shorter and shorter. Beginning of Winter is the start of winter and the coolness is replaced by the coldness when animals hibernate to pass the winter and plants decline. Winter Solstice indicates that the frigid winter formally begins.

According to temperature: Minor Heat, Major Heat, End of Heat, Minor Cold and Major Cold. These five Solar Terms are the representatives of the

《月曼清游图之踏雪寻诗》陈枚（清）
Chinese Painting: *Evening Tour: Making Poem on a Snowy Day*, by Chen Mei (Qing Dynasty, 1616-1911)

水量持续增加，为谷物带来生机；白露代表天气转凉，导致地面上的水汽凝结为露珠；寒露表示气温持续下降，露水快要凝结成霜；霜降表示天气寒冷使露水凝结为白霜；小雪表示严寒的天气使降雨减少，空气中的水分冻结成雪；大雪表示降雪量明显增加。

changing temperatures. Minor Heat reflects the weather becomes burning hot and Major Heat means the temperature reaches its annual peak. End of Heat implies that the burning weather draws to a close. The arrival of Minor Cold indicates the coming cold weather and Major Cold touches the lowest temperature of a year.

According to precipitation: Rain Water, Grain Rain, White Dew, Cold Dew, Frost's Descent, Minor Snow and Major Snow. The seven Solar Terms depict the changes in the annual rainfall. Rain Water marks the end of the dry and rainless weather in winter and the rainfall begins. Grain Rain brings vitality and White Dew stands for the moisture over the ground freezes into dewdrops according to the falling temperature. Cold Dew presents the dewdrops to freeze into frost due to the continuous temperature declination and Frost's Descent means the coldness turns the dewdrops into hoar frost. Minor Snow indicates that cold weather leads to decreasing rainfall and the moisture of the air freezes into the snow. And finally, Major Snow implies the obvious increasing snowfall.

According to phenology: Awaking of Insects, Pure Brightness, Grain Buds

物候类：惊蛰、清明、小满、芒种。这四个节气反映的是物候现象的变化。惊蛰时天气变暖，虫类结束冬眠，开始从土里移向地表活动；清明时天气晴朗，树木开始生长；小满形象生动地描述了夏季作物果实渐渐丰满、即将迎来收获的情形；芒种记录了小麦的成熟和谷黍类作物的播种。

我国古代劳动人民将二十四节气与物候相对应，还总结出了七十二候，即每一个节气分三候，每一候有五天。各候均有一种物候现象相应，称"候应"，分别有植物候应，包括植物的萌芽、开花、结果、衰落等；动物候应，包括动物蛰伏、生长、交配、迁徙等；非生物候应，如风雨雷电的出现与消失、始冻、解冻等。

and Grain in Ear, reflect the changes of the phonological phenomena. Awaking of Insects stands for the increasingly warmer weather and in this context, insects move to the Earth's surface. Pure Brightness indicates the clearness and prosperous growth of trees. Grain Buds vividly depicts the process by the summer crops become plump, namely the harvest is coming and Grain in Ear presents the maturity of wheat and marks the sowing of grain millet.

Ancient Chinese people related the Twenty-four Solar Terms with the corresponding phenological phenomena and even concluded the Seventy-two Pentads, namely one Solar Term divided into three pentads with five days for each pentad. Every pentad corresponds with a phenological phenomenon, known as *Houying* for plants including sprouts, blossoms, fruit-bearing and withering; for animals including dormancy, growth, mating and migration; and non-living things including the emergence and disappearance of wind, rain, thunder and lightning, as well as freeze and thaw.

> 二十四节气与中国人的生活

二十四节气是传统农事活动的重要依据，指导着人们按照自然规律，配合节气，从事相应的农事活

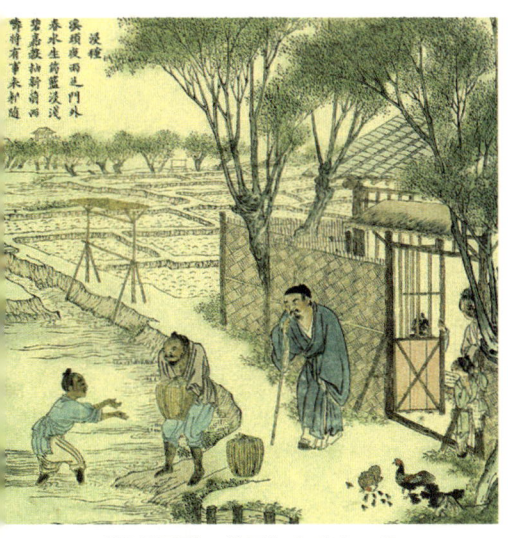

- 《耕织图册·耕图》焦秉贞（清）
Illustrated Book of Farming and Weaving: Ploughing, by Jiao Bingzhen (Qing Dynasty, 1616-1911)

> Twenty-four Solar Terms and Chinese People's Daily Life

The Twenty-four Solar Terms are the important guidelines to indicate people conduct corresponding agricultural activities according to the natural law and the characteristics of each Solar Term. Beginning of Spring marks the arrival of spring when all living things rouse up from sleeping and farmers begin to turn over the soil and apply fertilizer, preparing for the farming work. During the period after the Rain Water, the spring rain moistens the sleeping earth and people seize the precious opportunity and start to sow the seeds. After Awaking of Insects before of Spring Equinox, seeds germinate in light green and for southern China, people make full use of every minute to grow rice. Around Pure Brightness, melons and beans should be cultivated. For Grain Rain,

动。立春，万物萌动的春天来临，农民开始翻耕、施肥，为农耕做准备。随着雨水的来临，春雨开始滋润沉睡一冬的大地，人们开始抓紧时间播种。惊蛰过后、春分前后，种子露出嫩绿的枝芽，南方开始抢种水稻。清明前后，种瓜点豆。谷雨时节，早稻、玉米、棉花、花生等农作物开始春播。立夏，已经播种的作物开始进入生长期，人们忙着施肥、浇水、除草。小满，小麦等夏收作物的颗粒变得饱满。芒种，人们开始了夏种、夏收、夏

the early season rice, corn, cotton and peanut are available to be sowed in spring. Beginning of Summer implies that the planted crops begin to enter the growth period and people are engaged in applying fertilizer, watering and weeding. On Grain Buds, the summer crops like wheat are ripe. For Grain in Ear, summer planting, harvest and control are conducted by people and the hard farming work will continue until Summer Solstice. Farmers are busy with flood prevention and drainage as well as destroying harmful insects on Minor Heat and for Major Heat, after half a year's busy farming work, the harvest of early season rice comes. After Beginning of Autumn, the crops of southern China step into the harvest season, and the arrival of End of Heat means the crops of different regions, like corn, cotton and semilate rice generally become mature and during the harvest, people greet White Dew. Chinese people thresh and cultivate wheat in the period lasting from Autumn Equinox, Cold Dew to Frost's Descent. From Beginning of Winter to

• 河岸边的水车
Waterwheel on Riverside

- 北京日坛

日坛又叫"朝日坛",坐落在北京朝阳门外东南方的日坛路东,是明、清两代皇帝在春分日祭祀大明神(太阳)的地方。日坛坐东朝西,坛为圆形。

Temple of the Sun in Beijing

The Temple of the Sun, also the Altar of the Sun, located on the east of Ritan Road which is to the southeast of Chaoyang Gate in Beijing. It is the place for the emperors of the Ming and Qing dynasties (1368-1911) to offer sacrifices to the Sun, sitting toward the west with a round sacrificial altar.

管,直至夏至时节,农事活动都很繁重。小暑,农民忙着防洪、排涝、灭虫害。大暑,经过半年的忙碌,人们终于迎来了早稻的丰收。立秋之后,南方的作物进入丰收的季节。处暑来临,各地的农作物,如玉米、棉花、中稻等都进入了成熟期,人们在收获的忙碌中迎来了白露。从秋分、寒露直到霜降,人们都忙着打谷、种麦。立冬之后,直至大寒,在完成收割后,人们开始着手防冻、防虫等越冬准备工

Major Cold, people buckle to prevent frostbite and insects and conduct other preparations for winter after the harvest, which puts an end to the whole year's diligent cultivation.

Except for the application in agricultural areas, the Twenty-four Solar Terms are also tightly related to the daily life, festivals, traditional customs and religious ceremonies of Chinese people. Chinese people offer sacrifices to the Heaven, Earth, Sun, Moon and ancestors on Winter Solstice, Summer Solstice,

• 五谷杂粮有益健康
Coarse Cereals are Beneficial for Health

作，一年的辛勤耕作终于可以告一段落。

二十四节气除了指导农业生产外，与中国人的日常生活、节庆民俗、祭祀活动等也密不可分。冬至祭天、夏至祭地、春分祭日、秋分祭月、清明祭祖都是中国人格外重视的祭祀活动。除此之外，二十四节气对中医药文化、饮食文化也产生了重要的影响，如"冬病夏治""冬至饺子、夏至面""头伏饺子二伏面，三伏烙饼摊鸡蛋"等节气谚语，为人们健康生活提供了科学的依据。

Spring Equinox, Autumn Equinox and Pure Brightness respectively, regarded as important traditional sacrificial activities. In addition, the Twenty-four Solar Terms play an indispensable role in traditional Chinese medicine, food and culture. For instance, the diseases of winter should be treated in summer and the recommended food on certain Solar Terms like dumplings for Winter Solstice, noodles for Summer Solstice, as well as dumplings for the first period, noodles for the second period and Chinese baked pancakes with scrambled eggs for the third period of the triple-*Fu* (three hottest ten-day periods between Minor Heat and Beginning of Autumn), provide scientific basis for Chinese people's health lives.

二十四节气
Twenty-four Solar Terms

　　二十四节气的产生和发展反映了中国古代劳动人民的智慧和勤劳，千百年来，二十四节气一直影响和指导着中国人的生活。经过历朝历代的演化，逐渐形成了丰富多彩的节气文化。

The emergence and development of the Twenty-four Solar Terms reflect the wisdom and diligence of ancient Chinese people, and have been playing an important role in influencing and providing guidance for Chinese people's lives for thousands of years. After dynasties of evolution, the colourful culture of the Twenty-four Solar Terms has generally formed.

> 立春

　　立春是二十四节气中第一个节气，每年的2月3、4或5日，太阳到达黄经315°时为立春。所谓"立"，就是开始的意思，"春"即

- 枝条弯曲的龙游梅（图片提供：FOTOE）
Tortuous plum with twisted branches

> Beginning of Spring (*Lichun*)

Beginning of Spring is the first Solar Term of the year. On the 3rd, 4th or 5th of February, when the Sun travels to 315°celestial longitude is called as

春天，立春就是春天的开始。

根据元代吴澄的《月令七十二候集解》和清代曹仁虎的《七十二候考》等记载，中国古代将立春的十五天分为三候：一候东风解冻，大地气温有所回升；二候蛰虫始振，在冬季蛰伏的昆虫开始苏醒，蠢蠢欲动；三候鱼陟负冰，江河湖面上的冰开始融化，鱼开始从较为温暖的深水区上潜，在水面上随着破碎的冰片游动。

在民间有一种节气的计算方式叫做"数九"，就是从每年的冬至第二天开始算起，以九天为一个单位，从"一九"一直数到"九九"，共八十一天，用来记录天气从寒冷到和暖的过程。民间有谚语"春打六九头"，意思是说立春通常是在"五九"过后"六九"的第一天。立春过后，严冬开始进入尾声，气温回升，土地化冻。人们将立春当做春天的开始。

民谚有"一年之计在于春"的说法。自古以来，"春"就被中国人民重视，"立春"又称"打春""咬春""报春"，既是一个古老的节气，又是一个重大的节

Beginning of Spring. *Li*（立）means the beginning and *Chun*（春）means spring and therefore, *Lichun* marks the start of spring.

According to *Collection of Lunar Seventy-two Pentads* written by Wu Cheng in the Yuan Dynasty and *Study on Seventy-two Pentads* written by Cao Renhu in the Qing Dynasty, Beginning of Spring consists of 15 days and they are further divided into three pentads in ancient China: for the first pentad, the east wind unfreezes the earth and the temperature gradually increases; in the second pentad, insects wake up from the hibernation and start to move around; in the last pentad, lakes begin to thaw out and fishes can be clearly observed and move upward from the relatively warm and deep area, swimming with the dissolved ice on the water.

There is a popular method to calculate the Solar Terms known as *Shu Jiu* (counting the nine days) employed by Chinese people. Namely, from the 2nd day after Winter Solstice, there are nine periods of totally 81 days with each period consisting of nine days from the first to the ninth nine-day to record the process of the weather changing from coldness to warmness. It is said that the

日，千百年来流传下来许多风俗和仪式。

习俗

早在3000年前的周朝（前1046—前256），就有迎接"立春"的风俗。"芒神"为主管农事的春神。相传芒神居住在东方，所以立

- **句芒神**

句芒神，又名芒神、春神，主管农事之神，出自《山海经存》。

Goumangshen

Goumangshen, also known as *Mangshen* or the God of spring, is responsible for the agricultural activities, stemming from *Book of Mountains and Seas*.

spring starts from the sixth nine-day, indicating that Beginning of Spring is normally on the first day of the sixth nine-day, following the fifth nine-day. Beginning of Spring puts an end to the cold winter, when the temperature increases and soil thaws out. People treat Beginning of Spring as the start of spring.

The Chinese proverb prevails as making plans for the year in spring. Since ancient times, Chinese people have attached importance to spring and Beginning of Spring is also known as *Dachun* (Whipping Spring), *Yaochun* (Chewing Spring) and *Baochun* (Heralding Spring), which is not only an old Solar Term but also a significant festival, containing numerous customs and ceremonies of thousands of years' history.

Customs

As early as the Zhou Dynasty approximately 3,000 years ago (1046 B.C.-256 B.C.), ancient Chinese people obtained the custom of greeting Beginning of Spring. According to legend, *Mangshen* in charge of agricultural activities lives in the east and based on that, the emperor would lead the Three Councillors of State, Nine

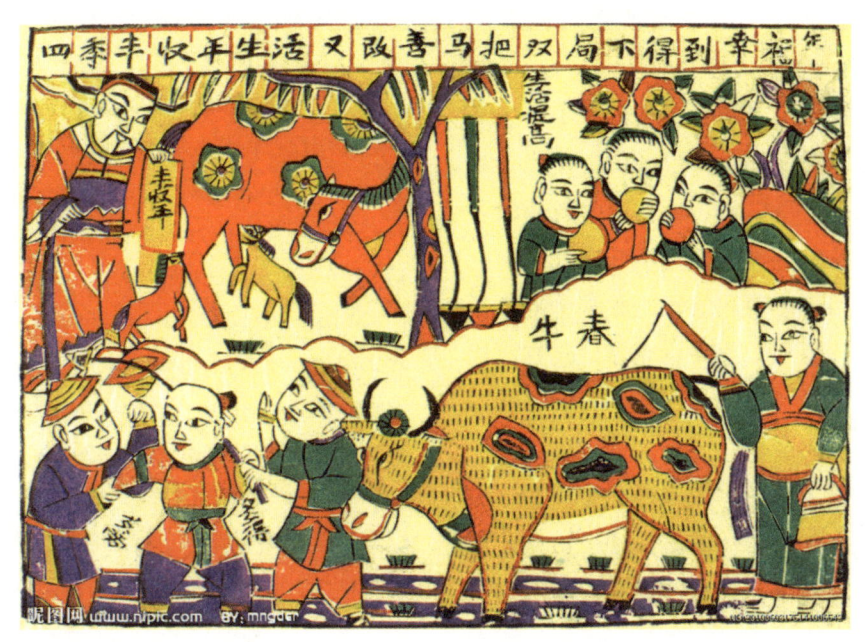

• 杨家埠木板年画《鞭春牛》
民间艺术家以年画、剪纸、彩塑等艺术形式，表现鞭春牛的习俗，深受百姓喜爱。
Yangjiabu Woodcut: *Whipping Spring Cattle*
Chinese folk artists use different artistic forms like the New Year picture, paper-cut and painted sculpture to illustrate the custom of whipping spring cattle, favoured by Chinese people.

春日，天子要率领三公九卿、诸侯大夫到郊外举行迎春祭祀仪式，祈求新一年农耕大丰收。宫廷、府衙门前也有迎春活动，活动内容随历史的发展也越来越丰富。到宋朝时，迎春活动发展成为官吏互拜的一种宫廷仪式，称为"拜春"。到了清朝，迎春从宫廷走向民间，成为全民参与的重要民俗活动。

Ministers, feudal princes and other senior officials to hold the sacrificial ceremony to greet the spring in the suburbs, in order to pray for the big harvest of a new year on the Beginning of Spring. The imperial court and feudal government offices also conducted activities to greet the spring with increasingly abundant contents. When it came to the Song Dynasty (960-1279), such activities developed to a royal

鞭春牛是一项古老的迎春习俗。立春又称"打春",这里的"打春"指的就是鞭打春牛。牛是古代农业的主要生产力,是农事的象征,所以自古就有以牛祝春的习俗。早在周朝,为了鼓励农耕,官府用泥、木和纸塑造成耕牛形象,称"春牛",并在立春日鞭打,相沿成俗。立春前一天,人们将装饰得喜庆鲜艳的春牛放置在城门外,称"立牛"。立春日那天人们轮流在春牛前作扶犁耕地状,边耕地边抽打春牛,载歌载舞,祈求丰年。春牛被打碎后,人们将碎片放置在家中,用以祈求丰收。

在立春这一天还有吃春盘、吃春饼、吃春卷、嚼萝卜的风俗,称为"咬春"。初春时节,吃一些新鲜蔬菜之类的食物,包含着喜迎春季,祈盼六畜茁壮、家业兴旺等吉祥寓意。唐代《四时宝镜》记载:"立春,食芦、春饼、生菜,号'春盘'。"意即把春饼与菜放在同一个盘子里。吃春盘的习俗可以追溯到晋代,到了唐宋时期,吃春盘、春饼的风俗盛行。宋代陈元靓在《岁时广记》中记载:"立春前一日,大内出春饼,并以酒赐近

ceremony for governmental officials to express appreciation, known as *Baichun* (Greeting Spring). In the Qing Dynasty (1616-1911), greeting the spring was popularised by normal Chinese people and became an important folk activity in which people participated.

Whipping the spring cattle is an old custom to greet the spring. Beginning of Spring is also known as *Dachun* (Whipping Spring), and *Da* is referred to whipping the cattle. As the symbol of agriculture, cattle played an indispensable role in agricultural productivity and therefore, it was also highly valued in the customs of greeting the spring in ancient China. In the early Zhou Dynasty, feudal official workshops employed mud, wood and paper to create cattle-liked statues, known as the spring cattle based on the purpose of encouraging farming. At the Beginning of Spring, people whipped the statues, which had been inherited as a custom. On the day before Beginning of Spring, ancient Chinese people placed the colourfully decorated spring cattle outside the city gate, regarded as *Liniu* (setting up the cattle) and on the following day, people took turns to make poses of ploughing and whipping the statue, and festively sang and danced to pray for a bumper harvest year. After breaking the statue, people would collect and place the fragments at home.

It is prevailing to eat spring rolls, spring pancakes and radishes at the Beginning of Spring, also called as the custom of *Yaochun* (Chewing Spring). During the first days of spring, eating fresh vegetables contains the auspicious meaning of greeting the spring and praying for thriving domestic animals and family property. According to *Treasure Mirror of Four Seasons* in the Tang Dynasty (618-907), at the Beginning of Spring, people should eat the spring dish, including radish, spring pancake and lettuce. Namely, the spring pancake and fresh vegetables are served in the same dish. This custom can be dated back to the Jin Dynasty (265-420) and in the Tang Dynasty (618-907) and Song Dynasty (960-1279), it was popular to eat the spring dish and pancake. Chen Yuanliang recorded it in *Records on Customs on Solar Terms* of the Song Dynasty: on the day before the Beginning of Spring, the imperial palace would cook spring pancakes and highly qualified liquor was granted to the courtiers. The emperor would grant

- 春饼

春饼又称"荷叶饼"，是一种温水烫面烙制或蒸制的薄饼。卷春饼的菜俗称"和菜"，生熟兼有，荤素齐全，除葱丝、甜面酱外，必有炒粉丝、豆芽、摊鸡蛋、炒韭菜，再配上豆腐干、酱肉等。吃春饼时，一定要卷成筒状，从头吃到尾，俗语叫"有头有尾"。

Spring Pancake

The spring pancake is also known as the lotus-leaf shaped pancake, a thin pancake cooked of baked or steamed dough made with warm water. The stuffing inside is called *Hecai* (accompanied dishes), consisting of fresh and cooked materials, rich in meat and vegetables, which includes shredded scallions, sweet flour sauce, fried bean vermicelli and sprouts, scrambled eggs and fried Chinese chives, accompanying with dried bean curd and braised pork with soy sauce. When eating the spring roll, it is particular about the shape of the roll and people should eat it from end to end, implying the meaning as "carrying a plan through to the end".

臣。"可见当时就有立春日吃春饼之俗，且多由皇上赐与近臣。明清时期，吃春盘、春饼之俗传向民间，清代《北平风俗类征·岁时》中记载了北京人家立春吃春饼的节俗，"立春，富家食春饼，备酱熏及炉烧盐腌各肉，并各色炒菜，如菠菜、韭菜、豆芽菜、干粉、鸡蛋等，且以面粉烙薄饼卷而食之"。

嚼萝卜也是咬春习俗。《燕京岁时记》中记载："是日，富家多食春饼，妇女等多买萝卜而食之，曰'咬春'。谓可以却春困也。"春卷也是立春食品，以面皮包馅炸制而成，多以猪肉、笋丝、豆芽、韭菜等为馅。

健康

立春正值冬天向春天过渡之时，万物萌动，阳气增长，天地之间呈现出一派欣欣向荣的景象。此时人体新陈代谢旺盛，正是调养身心的时候，尤其有利于肝脏的养护。但是由于天气乍暖还寒，早晚温差很大，而且经过一个冬天的蛰伏，人的免疫力低下，因此这个时节，要注意保健

the spring pancake and spring liquor to the officials and courtiers. In the Ming and Qing dynasties, eating spring dishes and spring pancakes was popularized the folk society and based on the records in *Records of Peiping Customs: Solar Terms* of the Qing Dynasty, people in Beijing obtained the custom of eating spring pancake on the Beginning of Spring, as "rich families eat spring pancakes and prepare braised pork with soy sauce, smoked meat, salted pork and various fried dishes like spinach, Chinese chives, bean sprouts, dry vermicelli and eggs, accompanying with baked thin pancakes made out of flour on the Beginning of Spring".

Chewing radish is another custom in this Solar Term. In *Book of Customs on Solar Terms in Beijing*, it is recorded that "on that day, rich people eat spring pancake and women buy and eat radish, known as *Yaochun*, which can help resist the spring drowsiness." The spring roll is also a representative food for the Beginning of Spring, made of flour crust fried in oil with the stuffing of pork, shredded bamboo shoots, bean sprouts and Chinese chives.

和预防疾病。在生活起居方面，人们强调"春捂秋冻"，注意保暖，不要贪图清爽而过早脱掉冬衣。另外，要早睡早起，积极参加户外活动，通过锻炼身体，提高免疫力，防止病菌的入侵。灿烂的阳光和盎然春意会让人心情愉

Health

The Beginning of Spring is the transitional period of winter and spring when all living things rise up from sleeping and the *Yang* spirit increases, presenting a thriving picture of the natural world. This is the time of vigorous metabolism, perfect for people to recuperate the body and mind, especially for liver maintenance. However, since the weather gets warmer suddenly and turns cold again, the temperatures in the morning and evening are significantly different. In addition, after a whole winter's hibernation, people's immune systems and disease resistance are comparatively weak and based on that, it is necessary to prevent disease and strengthen health care during this period. In terms of daily life, it is emphasized keeping warm based on the principle of being warm in spring and cool in autumn, without taking off their winter clothes too early to enjoy the coolness. What is more, people should

- 红枣枸杞茶
  Wolfberry Tea with Chinese Dates

• 《游春图》展子虔（隋）
Spring Outing, by Zhan Ziqian (Sui Dynasty, 581-618)

快，使身体保健取得事半功倍的效果。在饮食方面，要多甜少酸，有目的地根据时令变化和身体机制选择饮食。如大枣、花生、枸杞等有利于肝脏调理的食物都是此时不错的选择。

go to bed and get up early and take an active part in outdoor activities. Through physical exercises, immunity can be strengthened and diseases are easier to be effectively prevented. Meanwhile, the bright sunshine and the obvious message delivered by spring will create a cheerful mood, helping intensify the effect of health care. With regards to the diet, it is suitable to eat more sweet food than sour, and purposefully choose the right food according to the seasonal change and physical conditions. For instance, Chinese dates, peanuts and wolfberries are beneficial for the health of the liver.

> 雨水

雨水是二十四节气中的第二个节气，是反映降水现象的节气。每年的2月19日前后，太阳到达黄经330°为雨水。雨水一般从2月18日或19日开始，到3月4日或5日结束，即从"七九"的第六天到"九九"的第二天。

中国古代将雨水的十五天分为三候，第一候是獭祭鱼，即水獭开始捕鱼；第二候是鸿雁来，随着气温的回升，大雁从南方迁徙归来；第三候是草木萌动，大地在春雨的滋润下焕发出勃勃生机，植物开始抽芽生长。随着雨水节气的到来，冰雪纷纷融化，气温已经明显回升，降雪减少，降雨增多，有利于越冬作物的生长，故有"立春天渐暖，雨水送肥忙"的说法。雨水节

> Rain Water (*Yushui*)

Rain Water is the second Solar Term of the year, reflecting the rainfall. Around the 19th of February every year, when the Sun travelling to 330°celestial longitude is called the Rain Water, lasting from the 18th or 19th of February to the 4th or 5th of March, namely from the 6th day of the seventh nine-day to the 2nd day of the ninth nine-day.

Ancient Chinese people divided the 15 days of the Rain Water into three pentads: in the first pentad, the otter begins to hunt fishes; in the second pentad, with the rising temperature, wild geese return to the north from the south; in the third pentad, under the spring rain's moistening, all plants return to life and begin to sprout and grow. With the arrival of the Rain Water, the coldness starts to disappear. The temperature rises considerably with less snowfall

• 柳花
Willow Flowers

and more rainfall, which provides favourable conditions for the winter crops' growth. Therefore, there is a famous saying that day is increasingly warmer and the rain is busy bringing about nutrients to the soil and plants. The Rain Water is the best time for spring ploughing and people engage with the preparation work for the spring cultivation like seed selection, weeding and applying fertilizer.

Customs

In some areas in southern China, on the Rain Water, married daughters should visit their parents with presents like rattan chairs with red tape twined on them, for instance, to express the auspicious meaning of blessing parents' happiness and longevity. Another example is to send the stewed pettitoes with soybeans and sea-tangle in an earthenware cooking pot, sealed with red paper and string to present their gratitude for the love and care given by their parents. If it coincides with the daughter's wedding, the parents will send an umbrella to their daughter

气是春耕的最佳时机，人们开始选种、除草、施肥等春耕春播的准备工作。

习俗

在我国南方一些地区，雨水这一天，出嫁的女儿要带上礼物回娘家拜望父母。常送的礼物有藤椅，上面缠着红带子，有祝福父母福寿绵长的寓意；或者是用砂锅将猪脚、大豆和海带炖在一起，用红纸和红绳封口，以表示感谢父母的养育之恩。如果恰逢女儿新婚，父母要赠雨伞，意思是为女儿、女婿遮风挡雨，祝福他们在人生旅途上顺利平安。如果是久不怀孕的妇女，母亲会为其缝制一条红裤子，有求子之意。

"拉保保"是四川一些地区的民间习俗。所谓"保保"就是干爹的意思，为了让孩子健康顺利成长，父母要让孩子拜一位干爹，这一习俗流传至今。在雨水这天拜干

and son-in-law based on the meaning of keeping out wind and rain, and wishing them to be safe and smooth in the rest of their lives. For women who cannot get pregnant for a long time, their mothers will make red pants for the wish of having a baby soon.

La Baobao (getting sworn father) is a folk custom in the region of Sichuan Province. *Baobao* means the godfather or sworn father. Based on the purpose of ensuring the child can grow up healthily and stably, his or her birth parents will find the child's sworn father, which has long been regarded as a traditional custom. On the Rain Water, the children will visit their sworn fathers, so that they can obtain the grace from the rain and become easier to grow up. On the morning of Rain Water, the birth parents will prepare excellent liquor, delicious dishes and candles, and take their

- 四川广汉房湖公园举办的保保节

  （图片提供：FOTOE）

  广汉保保节，由民间传统"游百病"和"拉保保"的习俗演变而成。

  Baobao Festival in Fanghu Park in Guanghan City of Sichuan Province

  The *Baobao* Festival of Guanghan City is the evolution of the folk custom of *You Baibing* (walking away the disease, a physical exercise to eliminate diseases through walking) and *La Baobao*.

爹，取雨露滋润易生长之意。雨水日一早，父母就备好酒菜、蜡烛，带着孩子拜干爹。拜干爹时，要点上蜡烛，让孩子磕头行礼。

children to meet their sworn fathers. At the meeting, the birth parents will hold a banquet and light candles, and ask the children to kowtow to the sworn fathers.

元宵节

在雨水节气期间，中国的传统节日元宵节悄然而至。古人将"正月"称为"元月"，将夜晚称为"宵"，农历正月十五是一年中首个月圆之夜，因此称为"元宵节"。元宵节又称"上元节"，有一元复始的寓意。在这一天，家人共聚一堂，品尝元宵节的传统食物——元宵，并外出赏月，点燃焰火。元宵节标志性的活动是举办盛大的灯会，人们张灯结彩熙熙攘攘聚集在一起，观赏形态各异的灯笼，猜灯谜，欢度这个传统节日，因此元宵节又称"春灯节"。

• 吃元宵是元宵节的传统习俗
Eating Sweet Dumplings is the Traditional Custom of Lantern Festival

• 北京朝阳公园元宵节灯会（图片提供：全景正片）
Lantern Festival Show in Chaoyang Park in Beijing

Lantern Festival

During the period after the Rain Water's arrival, traditional Lantern Festival comes. Ancient Chinese people called the first month (*Zhengyue*) of the lunar year as *Yuanyue* and the night as *Xiao*. The 15th of the first lunar month obtains the first full moon night, known as the Lantern Festival or Shangyuan Festival, indicating the beginning of a new year. On that day, all family members get together and eat sweet dumplings as the traditional food representing this festival. At the same time, people admire the full moon outside and set fireworks. What is the landmark of this festival is to decorate with lanterns and streamers and hold magnificent lantern show, which will attract many people to admire various lanterns and guess lantern riddles. Based on that, the Lantern Festival is also called as the Spring Lantern Festival.

健康

雨水时节，寒冬向暖春过渡，天气变化多端。阳气持续增长，人们迎来了调和脾胃的最佳时期。但是这时天气变化无常，往往会使人的情绪产生波动，影响身心健康。另外，气温回暖也导致细菌、病毒滋生和传播。因此春季是各种疾病容易暴发的季节。

古人总结雨水时节气候反复、水量充沛的气候变化特点，"春夏养阳""春捂"等原则，遵循气候变化的规律，调整生活习惯。即在日常生活中注意保暖，预防感

• 萝卜养生茶
Radish Tea

Health

The cold winter is transferring to a warm spring after the Rain Water. The weather is changeable and the *Yang* spirit keeps growing, which is the best time for people to harmonize the spleen and stomach. However, people's emotions are likely to fluctuate influenced by the changeable weather, and produce negative effects on physical and psychological health. In addition, the increasing temperature will lead to the growth and spread of bacteria and viruses. Therefore, spring is the season of disease outbreaks.

Ancient Chinese people concluded that the Rain Water has changeable weather and abundant rainfall and based on that, has worked out the principles of tonifying *Yang* spirit in spring and summer and keeping warm in spring, satisfying the changing rules of the weather and providing effective suggestions on adjusting people's habits. It is suggested that people should pay attention to keeping warm, preventing colds, following a regular daily schedule, balancing the relationship between work and rest and strengthening physical exercise. In terms of diet, people should place emphasis on harmonizing the

• 水果中富含人体所需的维生素
Fruits are Full of Necessary Vitamins for Human

胃，同时还要注意规律作息，劳逸结合，锻炼身体。在饮食上，则侧重调养脾胃和祛风除湿。如果天气阴冷，可以适当进补，但要注意饮食清淡，少食油腻的食物，多吃新鲜蔬菜、多汁水果，如茼蒿、莲藕、韭菜、萝卜、柚子等，用以补充人体水分和维生素。

spleen and stomach, dispelling wind and eliminating dampness. If it is gloomy and cold, it is suitable to take a tonic, but the diet should be light without too much greasy food. Fresh vegetables and juicy fruits are the best choices to supplement body water and vitamins, like garland chrysanthemum, lotus root, Chinese chives, radish and grapefruit.

> 惊蛰

每年的3月5、6或7日，当太阳运行至黄经345°时为惊蛰。惊蛰是反映自然物候现象的节气。所谓"惊"就是惊动，"蛰"为藏，惊蛰即春雷乍动，惊动了严冬时节在洞穴里冬眠的动物和蛰伏在地下的昆虫。

惊蛰时节分为三候，第一候为桃始华，气温回暖，山里的桃花纷纷绽放；第二候为仓庚鸣，春风中绿树抽条，黄鹂立在树梢上鸣叫；第三候为鹰化为鸠，古人认为鹰会在惊蛰时化为布谷。春雷是惊蛰节气中最具有代表性的自然现象。人们习惯以惊蛰是否打雷来预测天气，有"未过惊蛰先打雷，四十九天云不开"的谚语。中国各地春雷初鸣的时间有所不同，南方大部分地区在惊蛰时可闻春雷初鸣，而华

> Awaking of Insects (*Jingzhe*)

On the 5th, 6th or 7th of March every year, when the Sun travelling to 345° celestial longitude is called the Awaking of Insects, reflecting the phenological phenomena. *Jing* means to wake up and *Zhe* is referred to hide, and their combination indicates that the spring thunder awakens all dormant animals and insects.

Three pentads constitute Awaking of Insects: in the first pentad, the peach blossom blooms in the mountains with the increasing temperature; in the second pentad, the green tree sprouts in the spring wind and oriole sings on the tree; in the last pentad, eagle turns into turtledove, a legendary male bird that would change into the cuckoo on the Awaking of Insects. The spring thunder is the most representative natural phenomenon of this Solar Term. People are accustomed to predicting the

梯田上耕种的农民
Farmers on Terraced Fields

weather situation through the spring thunder on the Awaking of Insects, and there is a saying "The thunder comes before the Awaking of Insects and after that, forty-nine days will be covered with clouds". The appearance of the first spring thunder in various regions in China differs and in the south part of China, most regions can hear the spring thunder on the Awaking of Insects. Compared with that, for southern China and the northwest regions, the spring thunder can be heard on the Pure Brightness. The Awaking of Insects is extremely important for farming and people have attached importance to this period since ancient times and regarded it as the beginning of the spring ploughing, known as the ceaseless spring cultivation on the Awaking of Insects. According to the poetry

南西北地区通常要到清明才能听到雷声。惊蛰节气在农事上有很重要的意义，人们自古就重视惊蛰，把它视作春耕开始的时节，有"到了惊蛰节，春耕不停歇"的说法。唐代韦应物的诗词《观田家》有"微雨众卉新，一雷惊蛰始。田家几日闲，耕种从此起"之句。此时华北地区冬小麦开始返青生长，但是土壤冷融交替，需要及时耙地来减少水分蒸发。江南地区的小麦已经拔节，油菜开始见花，应适时追肥，浇水灌溉。

习俗

古时候，虎患还是无法应对的灾难之一。相传在惊蛰时节，老虎会被响雷惊醒，出来作恶。广东一带的人们为了岁月平顺，就在惊蛰这一天拜祭白虎，以求万事顺利。

惊蛰雷动，蛇虫鼠蚁从泥土、洞穴中出来活动，祸害庄稼，传播病菌。古代没有农药，因此惊蛰时节，各地举行除虫仪式，希望赶走病害。广东客家人有在惊蛰这一天"炒虫"的风俗，用热水煮带皮的芋头，或炒豆子、炒米谷，象征害

A Farming Family of the Tang Dynasty (618-907) written by Wei Yingwu, four verses describe this Solar Term: The flowers bloom after the spring rain, and the spring thunder awakens all dormant insects. The leisure time for farmers runs out since the spring ploughing begins soon. At that time, the winter wheat in north China starts to turn green and grow. However, it is necessary to reduce the water evaporation in the soil through timely ploughing since the soil is thawing out. In the south of the lower reaches of the Yangtze River, the wheat has joined and the rape flowers can be seen. Farmers should apply sufficient fertilizer and irrigate the farm properly.

Customs

In Chinese folktale, the white tiger is the god of disputes and in ancient times, the tiger calamity was difficult to solve. It was said that on the Awaking of Insects, the white tiger would be awakened by thunder and did evils outside. People in the regions around Guangdong Province will worship the white tiger on this Solar Term for a smooth and peaceful time.

The thunder on the Awaking of Insects drives snakes, pests, mice and ants out of soil. They destroy crops

虫被吃了，寓意消灭虫害。山东地区则要在惊蛰这一天生火烙煎饼，寓意用烟火熏死害虫。而陕西等地则爆炒黄豆，象征将虫子炒熟吃掉。

and spread bacteria. In ancient times without pesticides, different regions held ceremonies to exterminate insects and expel viruses. The Hakka people in Guangdong Province obtain the custom of "frying insects" on the Awaking of Insects, and they boil taro without removing the peel, or fry beans, rice or grain on that day, meaning the insects are all eaten off and the insect attack is prevented. People in the Shandong region will bake pancakes on the Awaking of Insects, implying the meaning of suffocating the insects with smoke. And for Shaanxi Province and its surrounding region, people stir fry the soybeans, implying the insects are cooked and eaten by people.

- 白虎画像砖（西汉）
White Tiger Portrait Brick (Western Han Dynasty, 206 B.C.-25 A.D.)

健康

惊蛰时节，天气明显变暖，气候干燥，因此人们容易口干舌燥，咳嗽生痰。这时也是感冒、水痘等流行性传染病的易发期。所以人们要增强体质，提高身体免疫力，抵抗疾病入侵。在生活起居上，要保持室内湿度适宜，多开窗户通风，让空气流通；并且要早睡早起，保证充足的睡眠，要多做有益于身心的活动，如散步、慢跑、打球等，促进气血运行。在饮食上，应该清温平淡，多食用一些富含蛋白质、维生素的食物，如糯米、芝麻、蜂蜜、乳制品、蔬菜等，少吃燥烈辛辣的食物。

Health

On the Awaking of Insects, it becomes warmer significantly with an arid climate. Therefore, it is easy for people to have a parched mouth and scorched tongue, and cough with phlegm is common as well. Epidemics such as cold and varicella are likely to outbreak during this period. Therefore, people should strengthen physical exercises to enhance immunity and prevent illness. In terms of daily life, it is necessary to maintain comfortable indoor humidity and keep windows open regularly for ventilation. In addition, to ensure sufficient sleeping hours, people should go to bed and get up early and at the same time, spend more time strolling, jogging and playing ball games, which are beneficial for physical and psychological health and promote blood circulation. The diet during this period should be light and gentle, and protein-and vitamin-rich food is the best choice, like sticky rice, sesame, honey, dairy products and vegetables, without excessive strong and spicy food.

- 豆腐瘦肉煲

Tofu Casserole with Lean Pork

二月二

农历二月二是惊蛰前后的农历节日，民间认为这天是主管云雨的龙抬头的日子，流传着"二月二，龙抬头；大仓满，小仓流"的谚语。北方地区流行在这一天理发，叫"剃龙头"，寓意辞旧迎新，祈望好运。这天的民间饮食多以龙为名，如水饺叫"龙耳"，米饭叫"龙子"，馄饨叫"龙眼"。这些习俗寄托了人们祈求风调雨顺、五谷丰登的美好愿望。

Spring Dragon Festival

The second day of the second lunar month is the lunar festival around the Awaking of Insects and Chinese people believe this day is for the dragon to raise the head which is in charge of the clouds and rain, known as the Dragon Heads-raising Day on the 2nd of the second lunar month when "all granaries will be filled with crops and wheat" as quoted from the folk saying. People in northern China would like to have a haircut on that day, regarded as shaving the dragon's head and through doing so, people pray for good fortune as its implied meaning of ringing out the old and ringing in the new. The traditional food on that day is normally named after the dragon, such as the dumplings for that day called dragon's ears, rice as dragon's sons and wontons as dragon's eyes. These customs contain Chinese people's good wishes for favourable weather and a bumper grain harvest.

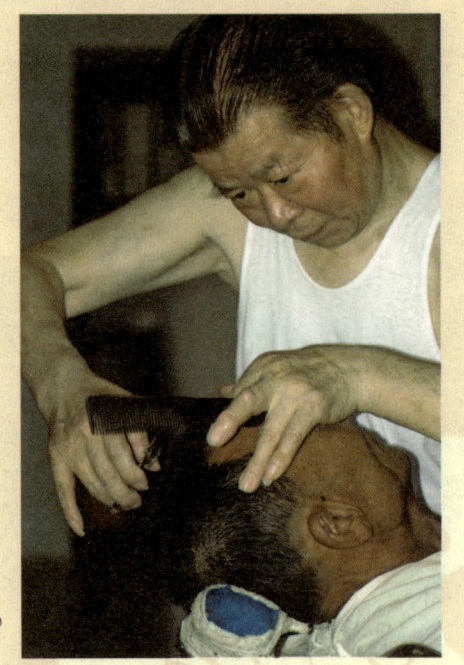

- 剃头的老人（图片提供：微图）
  Old Barber

> 春分

春分是反映四季变化的节气之一。每年3月20或21日,太阳行至黄经0°、日光直射赤道的日子为春分。所谓"分",是平分的意思,一是将一天平分为昼夜,各为12小时;二是春分正处于春季三

• 春花
Flower in Spring

> Spring Equinox (*Chunfen*)

Spring Equinox reflects the seasonal change and around the 20th or 21st of March every year, the Sun will travel to 0°celestial longitude and shine vertically on the equator. *Fen* means to equally divide and this obtains double meanings: it divides one day into daytime and night with each of 12 hours; meanwhile, as the Spring Equinox stays in the middle of the three months in spring and therefore, it also cuts the spring into two equal parts. Ancient Chinese people regarded Beginning of Spring, Summer, Autumn and Winter as the start of four seasons and Spring Equinox, Summer Solstice, Autumn Equinox and Winter Solstice as the middle of the four seasons. On the

• 江西婺源油菜花
Rape Flowers in Wuyuan City of Jiangxi Province

个月之中点，平分了春季。古代以立春、立夏、立秋、立冬来明确四季的开始，而春分、夏至、秋分、冬至则处于各季的中间。春分和秋分时阳光都直射赤道，昼夜时间相同，因此春分在古代又被称为"日中""日夜分""仲春之月"，民间也有"春分秋分，昼夜平分"的谚语。

Spring Equinox and Autumn Equinox, the Sun shines vertically on the equator while the daytime and night share the same length. Based on that, the Spring Equinox was called Midday, the equal dividing point of day and night and the moon of spring. According to folk wisdom, it is said that the Spring and Autumn Equinox equally divide the day and night.

春分分为三候，第一候是元鸟至，春分过后，燕子从南方迁徙到北方；第二候是雷乃发声，在下雨时风雨大作，雷声轰鸣；第三候是始电，下雨时打雷并伴有闪电。春分之后，白天变长，夜晚变短。春暖花开，莺飞草长，一派春意融融的景象。越冬农作物进入生长阶段，秧苗秀实，人们广兴水利，整地除害，精耕细作，施肥灌溉，有"春分麦起身，一刻值千金"的说法，用来形容这一宝贵的耕种时节。

习俗

在古代，春分是祭祀的重要节日，有"春分祭日，秋分祭月"的习俗。古代帝王的祭日场所设在东郊。祭日仪式非常隆重，有祭玉帛、奏礼乐、献祭品、行三跪九拜大礼等。春分这一天，百姓也举行隆重的民间祭祀活动，全村、全族人聚集在一起，在祠堂杀猪宰羊，奏乐歌舞，举行隆重的祭祖仪式，然后再分别祭奠各家的祖先。

送春牛也是春分的传统风俗之一。这一习俗源自周代，历代沿

The Spring Equinox is divided into three pentads: in the first pentad, after the Spring Equinox, swallows migrate from the south to the north; in the second pentad, the storm bursts and thunders roar; for the third pentad, lightning flashes while raining and thundering for the first time of a year. After the Spring Equinox, the day grows longer and the night becomes shorter. It is obvious to see the spring scenery as flowers blossom, orioles fly and grasses grow. The winter crops enter the growth phase and rice seedlings are growing well. Based on that, people concentrate on constructing water conservancy, soil preparation, deinsectization, intensive cultivation, applying fertilizer and irrigation, as a famous farmer saying goes: the wheat sprouts on the Spring Equinox and every minute counts, illustrating the precious time for cultivation.

Customs

In ancient times, the Spring Equinox is an important date for sacrificial activity, known as the custom of offering sacrifices to the Sun on the Spring Equinox and to the Moon on the Autumn Equinox. Ancient emperors built their places of worship in the eastern suburbs.

• 山西绛县年画《春牛图》
New Year Picture: *Spring Cattle* in Jiangxian County of Shanxi Province

The ancient sacrificial ceremony was magnificent and consisted of several steps as offering jade objects as an oblation, playing ritual music and giving grand salute of thrice kneeling and nine times bowing. On the Spring Equinox, normal Chinese people would hold abundant sacrificial activities, and the whole population of a village or a tribe got together, slaughtered pigs and sheep in the ancestral temple, and sang and danced happily. People treated it as an important ancestor-worshipping celebration, and they would hold the ceremonies for their ancestors after that.

Sending spring cattle is one of the traditional customs on the Spring Equinox, originating from the Zhou Dynasty (1046 B.C.-221 B.C.) and inherited by the following dynasties. According to the famous poet Lu You from the Southern Song Dynasty (1127-1279), the verses "the old man became bedridden for a long time, long enough to watch people sending spring cattle twice." vividly present the situation of sending spring cattle. In the early stage, the spring cattle was made of clay and when it came to the Qing Dynasty, the clay was replaced by paper. It was painted with patterns of cattle ploughing

袭。南宋诗人陆游有"老夫一卧三山下，两见城门送土牛"的诗句，描述的就是送春牛的景象。早期的春牛由泥土塑造而成，到了清代，泥牛改成用纸扎制，印有农夫牵牛耕田的图案，称为"春牛图"。春分日，人们挨家挨户赠送春牛图，还送上关于丰收的吉祥话，以表达祈盼五谷丰登的美好心愿。

另外，早在4000年前，人们就开始以竖鸡蛋来庆贺春天的来临。民间认为，春分这天可以很容易地把鸡蛋竖起来，有"春分到，蛋儿俏"的说法。竖鸡蛋要挑选光滑匀称的新鲜鸡蛋，此时的鸡蛋蛋黄下沉，鸡蛋的重心低，利于鸡蛋竖立。这一习俗流传至今。

健康

春分平分寒暑、昼夜，是一年之中气候较为温和的时节，此时雨水充沛，阳光明媚，草木生长繁茂，人体也处于新除代谢较为旺盛的时期，血液流动加快。但此时高血压、痔疮以及过敏性疾病都容易发作，天气的变化也容易引起腹泻、流感等疾病。因此人们应注意保持人体的阴阳调和。在生活起居上，要适当锻炼，规律睡眠，不要过分劳累，保持愉快的精神状态和平和的心态。在饮食方面应该少吃寒、热、刺激的食物，根据身体情况，保持膳食平衡，可以选择食用春笋、莴苣、豆苗、韭菜、桑葚、樱桃等时令蔬果。

led by a farmer, with important farming seasons and Solar Terms, regarded as the Painting of Spring Cattle. On the Spring Equinox, people send this picture from door to door with blessing words for bumper harvest to express their sincere hope for abundant harvest of all crops.

In addition, as early as 4,000 years ago, people celebrate the arrival of spring by erecting eggs. As Chinese people believe that, it is easy to hold up eggs on the Spring Equinox, known as "Spring Equinox Comes, Pretty Eggs Stand". It is necessary to choose new-born eggs of 4 to 5 days with smooth eggshells and in uniform size, based on the reason of the egg's low centre of gravity as the sinking yolk of the egg in this stage, which is beneficial for the egg to stand up. This is a custom passed down from the ancient times.

Health

As the Spring Equinox equally divides the cold and hot as well as daytime and night, it is regarded as the relatively mild period in a year, with abundant rainfall and bright sunshine, beneficial for the growth of grasses. During this period, human beings' metabolism is also in the comparatively vigorous stage, enjoying

• 新鲜的蔬菜
Fresh Vegetables

active blood circulation. However, high blood pressure, haemorrhoids and allergic diseases are likely to break out at the same time. The weather change during this period is the possible incentive for diarrhoea and influenza. Therefore, people should pay attention to regulating *Yin* and *Yang* spirits. With regard to daily life, appropriate physical exercises and regular sleeping without excessive fatigue are necessary. Meanwhile, it is also important to maintain a pleasant state and peaceful mind. In terms of the diet, cold, hot and other excitant food should be avoided and according to personal situation, people need to keep a balanced diet and choose seasonal fruits and vegetables like spring bamboo shoots, lettuce, bean seedlings, Chinese chives, mulberry and cherry.

> ## 清明

每年的4月4、5或6日，太阳运行至黄经15°时为清明。所谓"清明"，有天清地明之意。清代富察敦崇在《燕京岁时记》中引《岁时

> ## Pure Brightness (*Qingming*)

On the 4th, 5th or 6th of April every year, when the Sun travelling to 15° celestial longitude is the Pure Brightness. *Qingming* in Chinese means clear sky

《耕织图册·耕》焦秉贞（清）
Illustrated Book of Farming and Weaving: Ploughing, by Jiao Bingzhen (Qing Dynasty, 1616-1911)

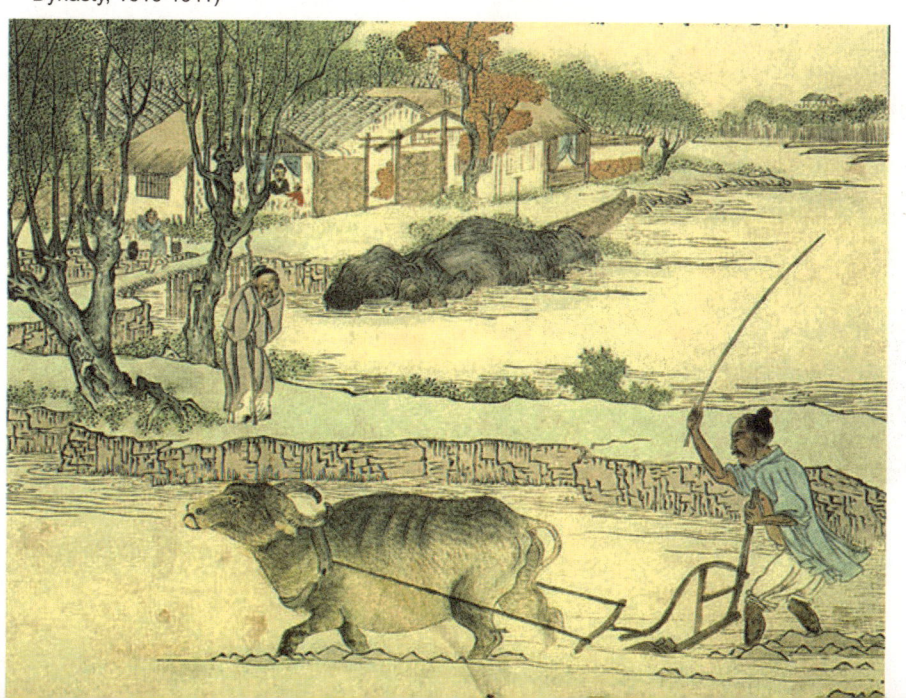

百问》："万物生长此时，皆清洁而明净。故谓之清明。"

古代将清明分为三候：第一候是桐始华，桐树开始抽枝生叶，白桐花迎来了花期；二候田鼠化为鴽，天气温暖，田鼠开始入洞，小鸟飞上枝头；三候虹始见，雨后的天空中可以看见彩虹了。清明时节，春意正浓，气温升高，正是春耕春种的大好时节，有"清明时节，麦长三节""清明前后，种瓜点豆"的民谚。

• 苏州网师园春景
Spring Scenery of Master-of-nets Garden in Suzhou City

and bright earth. Fucha Dunchong of the Qing Dynasty (1616-1911) wrote that all living things grow up during this period which is clean and bright, known as *Qingming*, in his Book *Customs on Solar Terms in Beijing* quoting from *Questions on Solar Terms*.

Ancient Chinese people divided the Pure Brightness into three pentads: in the first pentad, the tung tree begins to sprout and comes into leaf, and the tung flowers step into the flowering phase; in the second pentad, the field mice return to their burrows and the birds fly to the branches in the warm weather; for the third pentad, it is the first time in a year to see the rainbow when the Sun shines after the rain. During the period around the Pure Brightness, the spring scenery is charming. With the rising temperature, the wonderful opportunity for spring ploughing comes, reflected in the proverbs as the wheat grows significantly on the Pure Brightness, or melons and peas should be cultivated around the Pure Brightness.

The Pure Brightness is never limited to a Solar Term, but also a traditional festival for Chinese people to offer sacrifices to their ancestors. It originated from the feudal ritual of worshipping

- **江西婺源月亮湾**
Moon Bay in Wuyuan County of Jiangxi Province

清明不仅是节气，还是中国人纪念祖先的传统节日。清明节大约始于周代，源于古代帝王将相的"墓祭"之礼，后来传入民间，经历代沿袭，成为重要的祭祀节日。除了要禁火、扫墓祭祖之外，清明节还有许多丰富多彩的活动，如踏青郊游、放风筝、荡秋千、蹴鞠等民俗活动。

the ancient emperor's mausoleum, approximately early in the Zhou Dynasty (1046 B.C.-221 B.C.). After that, it was further popularized by the folk society and has been inherited and retained, regarded as an important sacrificial festival. The Pure Brightness festival includes abundant folk activities, like fire forbidding, tomb sweeping, spring outings, kite flying, swinging and Tzu-Chu (ancient Chinese soccer).

习俗

每当清明节，人们就会携带香烛纸钱、酒食果品去祭祖扫墓。人们把食物供奉在墓前，点燃香烛，

Customs

Chinese people still hold the tradition of sweeping the tomb with joss sticks and paper, candles, drinks, food and fruits to express their heart of worship towards

清明节的来历

最初清明节只是一个节气，但是由于清明与寒食节时间相近，于是这两个节日就合二为一了。唐代之前，寒食与清明是在内涵上有明显区别的两个节日。寒食节禁火，怀旧悼亡；而清明节祭亡，是为了佑生。相传春秋时期，晋文公重耳流亡期间，忠臣介子推追随保护，晋文公霸业大成之后，希望能够报恩，于是派人寻找介子推。介子推施恩不图回报，拒绝了晋文公的好意，并背着年迈的母亲躲进了绵山。晋文公派人上山搜寻，不见介子推的踪影，便下令放火烧山，希望借此逼介子推下山。谁知大火烧了三天三夜，仍不见介子推出来，大火熄灭后，人们发现他与母亲死在一

• 山西省绵山介子推墓（图片提供：FOTOE）
Tomb of Jie Zitui in Mianshan County of Shanxi Province

青团

在中国江南地区，人们有清明节吃青团的食俗。青团的做法是将糯米粉掺入菜汁揉成圆形，然后蒸熟。

Sweet Green Rice Balls

In the south of the lower reaches of Yangtze River, people retain the custom of eating sweet green rice ball on the Pure Brightness, which is steamed round sticky rice ball, made out of glutinous rice flour with green vegetable juice.

棵烧焦的柳树下。晋文公非常伤心，下令厚葬了介子推母子，在绵山建立祠堂，并将放火烧山的这一天定为寒食节，昭告天下，每年这天禁烟火，只吃冷食。

Origin of the Pure Brightness

The Pure Brightness obtained no other meaning than a Solar Term initially. However, people combined it with Cold Food Festival based on their similar dates. Before the Tang Dynasty (618-907), the Pure Brightness and Cold Food Festival shared considerably different connotations. On the Cold Food Festival, fire was banned to remember past times and mourn for the dead; compared with that, the sacrificial activity on the Pure Brightness was for blessing the living. It is said that in the Spring and Autumn Period (770 B.C.-476 B.C.), the Duke of State Jin, Chong'er was followed and protected by the loyal official Jie Zitui during his exile. When Chong'er accomplished the great undertakings, he hoped to repay the help and sent people to find Jie Zitui. However, Jie Zitui asked for nothing and refused the reward, and carried his old mother to Mianshan Mountain for hiding. After searching every corner of the mountain, Chong'er could still not find Jie Zitui and instructed people to set the mountain on fire, excepting to force Jie Zitui to appear. Three whole days' fire did not successfully achieve the goal and after putting out this fire, Jie Zitui and his mother died under a burnt willow. This grieved Chong'er deeply and he ordered people to bury Jie Zitui and his mother, and constructed a memorial temple on Mianshan Mountain. He further designated the day when the fire started as the Cold Food Festival, announcing that fire was forbidden and only cold food could be eaten on this day every year.

焚化纸钱，为坟墓培上新土，行作揖叩拜之礼，以表达对已逝亲人的思念哀悼之情。

　　清明时值晚春，春回大地，花红柳绿，正是踏青郊游的大好时节。中国古代就有清明踏青的习俗，人们呼朋唤友，举家出游，带着美食到郊外踏青，尽情享受春日的风光。唐代诗人杜甫的"三月三日天气新，长安水边多丽人"诗句，就描写了皇家游春踏青的胜景。宋代踏青之风盛行，北宋张择端的《清明上河图》也极其生动地

ancestor on the Pure Brightness. People offer food before the tomb, incinerate the joss paper, light candles, put fresh earth on and kowtow to the tomb, to express their condolence to the departed relatives.

　　The Pure Brightness is during the deep spring when spring returns to the Earth with blooming red flowers and emerald green willow, treated as the best time for spring outings and hiking. Since ancient times, these outdoor activities on the Pure Brightness have been prevailing. People invite family members and friends to go out with

- 《清明上河图》【局部】张择端（北宋）
Life along the Bian River at the Pure Brightness (Qingming) Festival, by Zhang Zeduan [Part] (Northern Song Dynasty, 960-1127)

柳树
Willow

delicious food in the suburbs, enjoying the charming spring scenery. Du Fu, a famous poet of the Tang Dynasty (618-907), described the spring outing of the imperial family as: It is bright and fresh on the 3rd of March, and many gorgeous women get together by the Qujiang River in Chang'an. This trend prevailed during the Song Dynasty (960-1279), vividly illustrated in the famous painting *Life along the Bian River at the Pure Brightness Festival* by Zhang Zeduan in the Northern Song Dynasty (960-1127).

On the Pure Brightness, people also fly various kites in spacious areas. By releasing the string in hand, the kite can fly high with the spring wind. On the one

● 风筝
Kite

描绘了人们踏青游春的热闹场景。

　　清明时节，人们还会带着各式各样的风筝来到宽阔的空地上，放开手中的线，让风筝乘着春风飞上高空。一方面，这时的气候非常适宜放风筝；另一方面，古时人们认为放风筝能够带走自己的晦气和噩运。因此人们待风筝升入空中后，往往将手中的线剪断，让风筝随风飘走，寓意消除病痛和灾难。

　　荡秋千也是我国清明节的传统游戏，早在南北朝时期就已经流行起来了。最初人们将彩带系在树枝

hand, the weather conditions is suitable for this activity; on the other hand, since ancient times, Chinese people have obtained the concept that flying kite could dispel misfortune. Therefore, after the kite flies high in the sky, people would usually cut the string and let the kite fly away, implying the meaning of eliminating disease and disaster.

　　Swing is also a traditional game in China, which prevailed early in the Southern and Northern dynasties (386 A.D.-589 A.D.). Initially, people tied coloured belts on the branches, and after years' evolution, it developed into the pedal swing with two ropes. In ancient times, women rarely went out, making the Pure Brightness Festival a precious opportunity for them to entertain and enjoy themselves. At present, playing on a swing is still favoured by Chinese people, especially for children.

Health

On the Pure Brightness, the weather is clear, bright and mild when plants germinate and apricot and peach flowers blossom. However, this period is also the time for the spring febrile diseases and various chronic diseases to outbreak and spread. Meanwhile, tomb-sweeping

《月曼清游图之杨柳荡千》陈枚（清）

Paintings of Scenes on Twelve Months in Imperial Palace of Qing Dynasty: Swing besides Willow Trees, by Chen Mei (Qing Dynasty, 1616-1911)

上，后来彩带发展成两根绳索加上踏板的秋千。古时候女性很少出门，所以清明节是一个难得的娱乐机会，可以尽情嬉闹玩耍。荡秋千至今仍为人们尤其是儿童所喜爱。

健康

清明正处于天清地明、气候温和、草木萌发、杏桃开花的时节，但是清明也是春瘟流行、各种慢性病多发的时期。清明节踏青扫墓，对亲人寄托哀思，容易触景生情，所以不宜过度悲伤，应尽快消

is to show people's grief to the departed relatives, but can easily ease their sadness. Based on that, it is inadvisable to wallow in the sadness excessively. People should make an effort to reduce the impact of negative emotion as soon as possible. In terms of daily life, it is important to keep a pleasant state of mind, get up and go to bed early and pay attention to indoor ventilation. In the enchanting spring sunshine, people can go to forests and parks to enjoy the fresh air and do physical exercise to relax the body. Although the custom of the Pure Brightness is

《品茶图》陈洪绶（明）
Painting of Tasting Tea, by Chen Hongshou (Ming Dynasty, 1368-1644)

除负面情绪的影响。在生活起居上，保持心情舒畅，早睡早起，注意室内通风换气。在明媚的春光里，可以选择去森林、公园等空气清新的地方，做一些舒缓的户外活动。在饮食上，虽然清明节有吃冷食的习俗，但是从健康的角度，还是以温热的食物为佳。在中医学上，此时养生以养肝为主，可以多吃一些荠菜、菠菜、山药等护肝养肺的食物，还可以食用白菜、萝卜、韭菜等蔬菜，对身体有益。此外，品饮明前茶，也有养肝清目、消除春困的功效。

eating cold food, however, it is better to have more tepid food with regards to the diet arrangement. According to Chinese medicine, nourishing the liver should be the focus during this period. For instance, vegetables like shepherd's purse, spinach and Chinese yam are beneficial for the liver and others like Chinese cabbage, radish and Chinese chives are also advisable. In addition, drinking the Mingqian tea (the green tea picked before the Pure Brightness) is also profitable for the liver and eyes, and obtains the effect of dispelling spring drowsiness.

明前茶

　　明前茶是清明前采摘加工的春茶。清明时节，寒冬退去，春暖花开，茶树经过漫长的冬季，积累了丰富的营养，长出了第一波茶芽，芽叶细嫩，口感香醇，有养肝明目、化痰除渴、提神醒脑的功效，是茶中佳品。由于能够达到采摘标准的产量极少，素有"明前茶，贵如金"之说。

• 西湖龙井茶
West Lake Longjing Tea

Mingqian Tea

Mingqian tea refers to the green tea picked before the Pure Brightness in spring. On the Pure Brightness, after a severe winter, the tea sprouts for the first time in a year when the tea shoots are delicate and tasty with the functions of reducing phlegm and refreshment, beneficial for the liver and eyes, known as the tea of high quality. Since the output of the tea satisfying the picking standard is limited, its value has long been compared with gold.

• 蒙山茶园
Tea Garden on Mount Mengshan

> 谷雨

谷雨是春季最后一个节气，每年的4月19至21日之间，太阳运行到达黄经30°时为谷雨。所谓谷雨，就是播种谷物、雨水增多的意思。明代王象晋的《群芳谱》中解释："谷雨，谷得雨而生也。"《月令七十二候集解》中说："三月中，自雨水后，土膏脉动，今又雨其谷于水也。盖谷以此时播种，自上而下也。"意思是说，三月中旬，雨水之后土地化冻，到了谷雨时节，寒潮天气基本结束，气温回升，雨水增多，十分有利于谷物生长。有农谚"雨生百谷"来形容此时降雨对于耕种的重要性。

谷雨时节分为三候：第一候为萍始生，雨水增多，气温回暖，地面水量有了明显增长，浮萍迎来了

> Grain Rain (*Guyu*)

Grain Rain is the last Solar Term in spring when the Sun travels to the 30° celestial longitude between the 19th and 21st of April every year. *Guyu* in Chinese means sowing grain and increasing rainfall. Wang Xiangjin explained its connotation in *Book of Flowers*: On the Grain Rain, the millet grows because of the rain. According to *Collection of Lunar Seventy-two Pentads*, it is recorded: After the Rain Water in the middle of the third lunar month, with the unfreezing of soil, the Grain Rain comes when the temperature and rainfall increase. It marks the end of the coldness of winter, which is beneficial for the grain growth. The famous saying among Chinese farmers: rain breeds cereals, reflects the importance of rain during this period of cultivation.

生长期；第二候为鸣鸠拂其羽，鸣鸠即布谷鸟，在树上鸣叫，提醒人们开始播种；第三候为戴胜降于桑，在桑树上可以看到戴胜鸟了。谷雨时节，秧苗初插，作物新种，最需要雨水的滋润，我国南方大部分地区雨水充沛，对水稻栽插、玉米生长都非常有利。

The Grain Rain is divided into three pentads: in the first pentad, with the increasing rainfall and temperature, the water volume above ground grows significantly and the duckweed enters its growth phase; in the second pentad, the cuckoo sings on the tree, reminding people to sow; for the third pentad, the Eurasian Hoopoe can be seen on the mulberry. On the Grain Rain, the rice seedlings are newly transplanted and need the moisture from the rain most. The abundant rainfall in most regions of southern China is plentiful, in favour of the growth of rice and corn.

- 安徽黄山毛峰茶园
  Maofeng Tea Garden on Huangshan Mountain in Anhui Province

关于谷雨的传说

相传五千多年前，轩辕黄帝的左史官仓颉游历名山大川，上观星辰云雨，下观山川河海、飞禽走兽，创造出了最早的象形文字。"仓颉造字，而天雨粟"，上天降下谷子雨以奖励仓颉的功绩。于是人们便把这一天称为"谷雨"，称仓颉为"文字始祖"，每年这天祭祀仓颉。

Legend of Grain Rain

According to legend, more than 5,000 years ago, Cangjie, the official historian of the Emperor Huang, travelled around well-known mountains and rivers, observed stars, clouds and rain in the sky and conducted researches on the mountain, river, ocean, fowl and beast on the ground. Based on his experience and knowledge, Cangjie created the earliest hieroglyphic. It was said that after Cangjie successfully produced character, it rained grains from the sky in order to reward Cangjie for his great contribution. Therefore, people call this day as the Grain Rain and regard Cangjie as the Ancestor of Chinese Character. At present, Chinese people worship Cangjie on this day every year.

- 仓颉造字

Invention of Chinese Characters by Cangjie

习俗

谷雨不仅是农事的重要时节，还是渔民出海捕捞的好时机。谷雨后春江水暖，鱼经过严冬的蛰伏开始在浅海区域活动，此时适合出海捕捞。为了祈求出海平安、鱼虾满

Customs

The Grain Rain not only plays an indispensable role in guiding agricultural activities, but also provides a precious opportunity for fishermen to start their fishing operations. During this period, the weather becomes warmer, which allows fish to swim around in the neritic region after a long period in cold winter, known as the wonderful opportunity for fishing. In order to pray for the safety of the operations and a good harvest in fishing, people living in the coastal regions maintain the custom of offering sacrifices to the sea on the Grain Rain. On that

- **福建西陂天后宫供奉的妈祖像**

  妈祖又称"天妃""天后""娘妈"。传说她能乘席渡海，预测天气变化，救助过许多渔舟商船，民间称她为"神女""龙女"。因感其救世济人的恩惠，沿海人们尊她为海神，立庙祭祀。历代渔民和航海者在起航前都要先祭妈祖，祈求平安，在船上还供奉妈祖神位。

  Statue of Goddess Matsu in Tianhou Temple in Xibei County of Fujian Province

  Matsu is also known as the Celestial Queen or Sea Goddess. It is said that she can cross sea on a flying blanket, predict the weather change and had saved numerous fishermen and merchant ships. Ancient Chinese people respected her as Goddess or Dragon Lady. Based on the gratitude to her grace, people living in the coastal region build temple to worship her. Previous fishermen and voyagers would offer sacrifices to Matsu before departure, praying for safety, and place her statue on the ship.

仓，在沿海地区，形成了在谷雨这一天祭海的习俗。渔民们在良辰吉时举行盛大的祭祀仪式，因此，谷雨这天也被称为渔民出海捕鱼的"壮行节"。

在山东、陕西、山西一带，民间有谷雨时节除灭五毒的习俗。所谓"五毒"，是指蝎子、蛇、壁虎、蜈蚣和蟾蜍。谷雨时天气潮湿温暖，正是虫害频发的时节，人们在此时制作年画，绘制五毒图案，

day, fishermen conduct the magnificent sacrificial ceremony on the most auspicious occasion and therefore, the Grain Rain is also called the Departure Festival.

In the regions of Shandong, Shaanxi and Shanxi provinces, the local custom of destroying five poisonous creatures is prevailing. The five poisonous creatures refer to scorpion, snake, gecko, centipede and toad. Since the weather of the Grain Rain is humid and warm when insect

• **肚兜上的虎食五毒图案**

五毒是指蝎子、蛇、壁虎、蜈蚣、蟾蜍五种害虫。民间在衣饰上绣制五毒，或是在饼上画五毒图案，还有的用纸剪出五毒图案，贴在门、窗、墙、炕上，均有驱除害虫之意。

Pattern of Five Poisonous Creatures on Bellyband

The five poisonous creatures are scorpion, snake, gecko, centipede and toad. Chinese people apply their patterns on clothes, Chinese cakes and paper-cut. People paste the paper-cut with their patterns on door, window, wall or the heatable brick bed, implying the meaning of destroying pests.

• 牡丹花

牡丹是中国特有的名贵花卉，花朵硕大，花色绚丽，雍容华贵，有"花王"之称。

Peony

Peony is an exclusive precious flower in China, with large petals and bright colour, which is dignified and graceful, known as the King of Flowers.

称作"谷雨贴"，以寄托除害虫、祈盼平安与丰收的美好愿望。

牡丹是中国特有的一种观赏性花卉，颜色鲜艳，造型丰富多变，有着雍容华贵的气度和端庄秀丽的神韵，素有"百花之王"的美称。牡丹花的故乡在河南洛阳，每到谷雨时节，牡丹花便会盛开，因此牡丹花又被称为"谷雨花"，并演化出谷雨赏牡丹的习俗。

唐代诗人刘禹锡就曾作诗《赏牡丹》，细致入微地描绘了当时京城洛阳的胜景："庭前芍药妖

attack is likely to happen, people wipe out destructive insects and draw the New Year picture with the patterns of five poisonous creatures, known as the Grain Rain picture, wishing for no insect attack, safety and a bumper harvest.

The Peony is a Chinese-specific ornamental flower that has bright colour and a gorgeous profile, known as the symbol of elegance and dignity and enjoys the reputation of King of Flowers in China. The origin of the peony is Luoyang City in Henan Province. Peony blossoms during the period around the Grain Rain and based on that, Chinese people call her as the Grain Rain Flower and obtain the custom of admiring peony on the Grain Rain.

In the poem Admiring Peony written by Liu Yuxi of the Tang Dynasty (618-907), he precisely illustrated the magnificent scene of the capital city

• 《簪花仕女图》周昉（唐）

仕女高高的发髻上簪一朵牡丹花，侧身右倾，左手执拂尘引逗小狗。

Court Ladies Adornig Their Hair with Flowers, by Zhou Fang (Tang Dynasty, 618-907)

The graceful hair bun of the court lady on the right is decorated with a peony flower. She leans to the right and holds a horsetail whisk in her left hand, playing with a puppy.

无格，池上芙蕖净少情。唯有牡丹真国色，花开时节动京城。"意思是说庭前的芍药妖艳无比却少格调，池中的荷花虽然清净却少感情，只有牡丹才是真正的天姿国色，在花开之时惊动了整个京城的人前来观赏。宋朝之后，洛阳牡丹更是享誉天下，观赏牡丹的风俗历久不衰，时至今日，人们依然在谷雨时节相约在洛阳，一睹牡丹的风采。

Luoyang at that time: The gorgeous peonies are growing in the yard. Compared with peony, lotus is pretty but lacks of interest. Only a peony obtains the real beauty and the whole capital city will be touched by its blossom, attracting all people to admire. After the Song Dynasty (960-1279), the peony was well known in the world and the custom of admiring the peony has been maintained until now. At present, people get together in Luoyang City and admire peonies on the Grain Rain.

健康

谷雨是春季的最后一个节气，气候温暖、多雨潮湿是这个时节最显著的气候特征。人们应该遵循节气的变化，随时增减衣服，避免或减少疾病发生。另外，由于空气中湿度逐渐加大，此时也是关节肿痛等病症的高发期，应该注意防潮保暖，避免着凉。在饮食方面，谷雨时节有着流传甚广的习俗，并且南北分明。北方谷雨时吃香椿，可以健胃理气，提高人体的免疫力；南方采制谷雨茶，安心益气，清肝明目。

Health

The Grain Rain is the last Solar Term of spring when the weather is warm and humid with abundant rainfall. People should pay attention to their clothes-choosing based on the purpose of following the weather change during this period to prevent diseases. In addition, since the air humidity increases gradually, this period is also a threshold of joint pain, and protection against dampness and keeping warm should be highlighted to avoid the possibility of catching a cold. In terms of diet, the customs of the north and south of China are considerably different with their own traditional and widespread practices. On the Grain Rain, people in northern China eat tender leaves of Chinese toon that is beneficial for the stomach, helps regulate vital energy and strengthens the immunity system of the human body; people in southern China drink Guyu tea, which makes one feel at ease and energetic, clears the liver and improves vision.

• 香椿炒鸡蛋（图片提供：微图）
Scrambled Eggs with Tender Leaves of Chinese Toon

谷雨茶

　　谷雨茶，就是在谷雨时节采制的茶叶，又称"二春茶""雨前茶"，与明前茶一样，都是茶中佳品。谷雨时节，气候温暖，雨量充沛，茶树生长旺盛。茶树经过冬季的休养，此时长出的芽叶饱满鲜嫩，色泽翠绿，醇香宜人。谷雨茶不仅口感好，而且营养丰富，富含多种维生素和氨基酸，具有清火、明目、除湿气等功效，所以，谷雨茶受到茶人的追捧。

Guyu Tea

Guyu tea is the tea picked during the period of the Grain Rain, also known as mid-spring tea or Yuqian tea (tea picked after the Pure Brightness and before the Grain Rain), sharing the same fame with the Mingqian tea. The weather is warm and rainfall is plentiful on the Grain Rain when the growth of tea trees is prosperous. After a whole winter's dormancy, the newly growing leaves are fresh and delicate in emerald green with a pleasant aroma. Guyu tea is tasty and contains abundant nutrition, rich in various vitamins and amino acids. Drinking Guyu tea can clear away heat, improve eyesight and remove dampness, favoured by Chinese people.

- 饮茶是一种健康的生活方式
  Tea-drinking is a Healthy Habit

> 立夏

　　每年的5月5、6或7日，太阳到达黄经45°的日子为立夏。《月令七十二候集解》中记载："立，建始也""夏，假也，物至此时皆假

> Beginning of Summer (*Lixia*)

On the 5th, 6th or 7th of May every year when the Sun travels to 45° celestial longitude, this day is called the Beginning of Summer. According to *Collection of Lunar Seventy-two Pentads*,

- 江苏宜兴竹海
Bamboo Sea in Yixing City of Jiangsu Province

大也"，意思是说立夏是季节转换的节气，标志夏天的开始。在这个节气里，春天播种的万物都已经长大了。

立夏节气分为三候：第一候蝼蝈鸣，夏天到了，田里蛙声一片；二候蚯蚓出，蚯蚓从湿润柔软的土地中钻爬出来，在田地里留下了湿漉漉的痕迹；三候王瓜生，王瓜的枝蔓开始快速攀爬生长。

Li (立) refers to the start and *Xia* (夏) means summer and during this period, all living things are growing vigorously. It indicates that the Beginning of Summer is the symbol of seasonal transformation, marking the start of summer. The crops planted in spring grow up on this Solar Term.

The Beginning of Summer is also classified into three pentads: in the first pentad, the croaking of frogs and coos of mole crickets can be clearly heard in the field; in the second pentad, earthworms crawl out from the underground with a damp trace behind them; for the third pentad, the vines of snake gourd develop quickly and climb high.

The Beginning of Summer is the turning point of spring and summer, with significant temperature increases and the natural world enters the prosperous phase at this time, described in *Liansheng Baqiang* written in the Ming Dynasty: During the period around the Beginning of Summer, the season changes when all living things are energetic. Farmers step into the busy season when they should transplant the early seasonal rice

- 鱼戏莲叶间
  Fish Swimming through Lotus Leaves

立夏是春夏交替的时节，此时气温明显升高，明人《莲生八戕》中以"孟夏之日，天地始交，万物并秀"来描述立夏的繁茂景象。农民进入农忙时节，早稻插秧，中稻播种，夏收的农作物已经准备进入丰收季节。

习俗

古人对立夏非常重视，周朝时，立夏这天帝王要身穿朱色礼服，乘坐朱色马车，高举朱色旗帜，率领文武百官到郊外举行迎夏仪式，祭祀炎帝和祝融。相传炎帝和祝融在远古时期共同统治南方。炎帝是太阳之神、农业之神，而祝融是火神。中国自古以来重视春耕夏耘，太阳是影响农作物生长的主要因素，所以，帝王在立夏日到郊外举行隆重的祭祀仪式，以祈求风调雨顺、五谷丰登。

立夏日各地还有许多有趣的风俗，较为普遍的是秤人习俗。这个习俗起源于三国时期，南方少数民族首领之一孟获被蜀国丞相诸葛亮收服，归顺蜀国。诸葛亮临终之际，不放心年幼的蜀主阿斗，于

seedlings, sow the seeds of middle-season rice. At the same time, people are well-prepared for the harvest of the summer crops.

Customs

Ancient Chinese people attached importance to the Beginning of Summer. From the Zhou Dynasty (1046 B.C.-221 B.C.), the imperial emperor would wear a red ceremonial robe and ride in a red carriage with red flags held high, leading all the civil and military officials to conduct a grand ceremony in the suburb to greet the arrival of summer. In this ceremony, people would offer sacrifices to Emperor Yan and the God of Fire. According to legend, Emperor Yan, the God of the Sun and Agriculture, and the God of Fire dominated the southern region jointly in remote ancient times. The spring ploughing and summer cultivation have been thought highly of since ancient times. The Sun is the main factor that influences the growth of plants and therefore, ancient Chinese emperors prayed for good weather for the crops and bumper grain harvest by holding the summer sacrificial ceremony in the suburbs.

The customs at the Beginning of

• 北京先农坛观耕台

先农坛位于北京南郊,是明、清两代皇家祭祀先农诸神的场所。观耕台是古代皇帝亲耕完毕、观看文武百官耕作的高台。

Cultivation Observation Platform in Temple of Agriculture in Beijing

The Temple of Agriculture is located in the southern suburbs of Beijing, known as the place for the imperial families in the Ming and Qing dynasties (1368-1911) to worship the Gods of Agriculture. The Cultivation Observation Platform used to be the place for the emperor to watch the civil and military officials do farm work after his inspection.

是嘱咐孟获每年都要来看望一次。孟获每年立夏这天都依诺前来看望阿斗,每次去看望的时候都要称一下阿斗的体重。后来立夏秤人的习俗流传下来,成为人们祈求清净安乐、福寿双全的活动。每逢立夏日,人们就支起一杆大秤,秤钩悬一个凳子,轮流坐到凳子上面秤重。司秤人一面打秤花,一面讲着吉利话,秤小孩就说"秤花一打二十三,小官人长大会出山。七品

Summer are interesting and vary from place to place, among which the most common is weighing people. This custom originated from the Three Kingdoms Period (220-280) when, one of the chieftains of the southern ethnic groups, Meng Huo was subdued by Zhuge Liang, the prime minister of State Shu. Before Zhuge Liang died, he felt worried about the young leader of State Shu Adou and committed him to Meng Huo, asking Meng Huo to visit Adou

县官勿犯难，三公九卿也好攀"，秤老人则说"秤花八十七，活到九十一"。

在立夏这天，还有吃"立夏蛋"的习俗。民间认为，鸡蛋寓意圆满，有"立夏吃了蛋，热天不疰夏（疰夏是种夏季常见的病症，通常表现为腹胀厌食、消瘦乏力）"的说法。立夏前一天，人们用茶叶

- 立夏秤人的风俗（图片提供：FOTOE）
Custom of Weighing People on Beginning of Summer

once a year. According to his promise to Zhuge Liang, Meng Huo went to see Adou at the Beginning of Summer every year. Meng Huo weighed Adou whenever he met with Adou. Since then, weighing people has become a custom until now, known as the activity for people to pray for a peaceful life, good fortune and longevity. On the Beginning of Summer, a big weighing scale connected with a stool will be prepared in advance, and people are weighed in turns by sitting on the stool. The person who is responsible for weighing reads the weight and says auspicious words. When a child is weighed, he says: when sitting on the weighing scale with the pointer indicating the mark 23, the child will obtain a wonderful future. To be the county magistrate is never a difficult task and the Three Councillors of State and Nine Ministers would possibly be the patronage. When an old person is weighed, he says: when sitting on the weighing scale with the pointer indicating the mark 87, you will live through 91.

At the Beginning of Summer, it is also prevailing to eat summer eggs as a traditional custom. Chinese people believe that an egg obtains auspicious meaning based on its round shape, known as "Eating the summer egg can prevent people from summer diseases (refer to abdominal distension, inappetence, emaciation and lack of strength)". On the day before the Beginning of Summer, people boil eggs in the tea dust soup with walnut shell and other materials, and let the children eat the eggs, praying for the health, safety and growth of children. At present, eating summer eggs is not limited to a convention at the Beginning of Summer. The summer eggs, known as tea-flavoured eggs, are favoured by Chinese people.

Another custom about eggs in this Solar Term is called egg game, one of the favourite games for Chinese children. At the Beginning of Summer, parents put the boiled eggs into the mesh bag after immersing them in cool water and children will wear the bag around their necks. Then, they would get together to knots and begin the egg game. The egg has two ends with the pointed one at the top and the round one at the bottom. When the game starts, one person holds two eggs in each hand, and uses the top to hit the bottom or vice versa. According to the rule, the person whose egg is broken in this process would be the loser.

● 茶叶蛋（图片提供：微图）
Tea Flavoured Egg

末、胡桃壳等材料将鸡蛋煮熟，给家中的孩子吃，祈求孩子健健康康，平安成长。现在立夏蛋不再拘泥于立夏日食用，发展成为传统的小吃——茶叶蛋。

除了吃蛋，还有深受孩子喜爱的斗蛋习俗。立夏日，家家户户把煮好的鸡蛋用水浸凉之后，装进丝网袋，挂在孩子的颈上。孩子们便三五成群，玩起斗蛋游戏。蛋分两端，尖者为头，圆者为尾。斗蛋时，一人手持一只蛋，用蛋头击蛋头，蛋尾击蛋尾，蛋壳先破者认输。

健康

立夏时节，正是春夏交替的季节，天气渐渐炎热，人体新陈代谢旺盛，要顺应节气的变化，特别要注意保养心脏。年轻人应该强身健体，控制自己的情绪，达到以心养身的目的。老年人则应该避免气血瘀滞，以防心脏病发作。在生活起居上，因天气逐渐炎热，昼长夜短更加明显，应适当午睡，养成良好的生活作息。要保持心情舒畅，切忌大喜大怒，保持充足的体力和饱满的精神状态。立夏后气温升高，

Health

The Beginning of Summer is the turning point of spring and summer when the weather becomes hotter and people have active metabolism. It is necessary to adapt to the season change and the heart should be especially paid attention to. For the young, it is important to do more physical exercises and control one's emotions to attain the goal of preserving health. For the old, one should avoid the stasis of blood and *Qi* in case a heart attack occurs. In terms of daily life, it is more obvious to feel the increasing temperature and see the longer days and

- 西湖边练剑的老人
  Old People Practicing Taiji Sword by West Lake

人容易出汗，还要注意运动适度，可以选择相对平和的活动，如打太极拳、舞太极剑、散步、钓鱼、练书法、绘画、下棋等。在饮食上，应以少盐、少脂、清淡、富含维生素为主，多食用瓜果蔬菜和粗粮。

shorter nights. Based on that, people should develop a healthy and scientific daily schedule. For instance, afternoon naps play a positive role in people's health. Meanwhile, maintaining a good mood, avoiding severe emotional fluctuation, and keeping physical strength and a stable mental state is advisable. With the arrival of the Beginning of Summer, the temperature rises and it is easy to perspire. Therefore, the amount of exercise should be moderate and it is better to choose mild activities, like playing Taijiquan (a traditional Chinese shadow boxing), playing Taiji sword, strolling, fishing, practicing calligraphy, painting and playing chess. With regards to the diet, the food should not be salty and greasy, but light and rich in vitamins, and vegetables, fruits and coarse grains are worth recommending.

- 练习书法可以修身养性
  Practicing Calligraphy Can Cultivate Mind

> 小满

每年的5月20、21或22日，太阳运行到黄经60°的日子为小满。小满一词，在南方和北方有着不同的含义。在北方，"满"是指夏

> Grain Buds (*Xiaoman*)

Grain Buds is on the 20th, 21st or 22nd of May every year when the Sun travels to the 60° celestial longitude. This Solar Term refers to different meanings in southern and northern China. In

- 安徽婺源水田
Paddy Field in Wuyuan County in Anhui Province

熟作物籽粒的饱满程度。古代历书记载："斗指甲为小满，万物长于此少得盈满，麦至此方小满而未熟，故名也。"意思是说，在小满时节，大麦、冬小麦等夏收作物已经灌浆结果，籽粒渐渐饱满，但尚未成熟，所以叫小满。而在南方，"满"是指雨水的丰盈程度。农谚"大落大满，小落小满"，意思是说雨水越充沛，越是丰收的征兆。

小满分为三候，第一候苦菜秀，在小满时节，苦菜已经生长得非常繁茂了；第二候靡草死，一些细软的野菜在炎热干燥的气候和阳光强烈照射下开始枯死；第三候麦秋至，麦子结出沉甸甸的麦穗，进入成熟期。从农业上讲，小满节气正处于盛夏时分，是夏收、夏种、夏管交汇的季节，因此小满是农民一年中最为繁忙的季节。

习俗

在浙江海宁一带，小满时节有抢水的习俗。所谓抢水，就是一种将水引灌入农田的仪式，意在祈求雨水充足、五谷丰登。在南方，此时正是早稻追肥、中稻插秧的重要

the northern part of China, *Man*（满）indicates the extent of plumpness of grain with regards to the summer crops. According to the ancient almanac, on the Grain Buds, barley and winter wheat have finished the filling stage and bore fruits, with the kernels becoming plumper but not ripe, known as *Xiaoman*. Compared with that, in southern China, *Man* means the volume of rainfall, based on the agricultural proverb that: the more the rainfall is, the bumper the harvest will be.

The Grain Buds is divided into three pentads: in the first pentad, the bitter herbs have enjoyed prosperous growth; in the second pentad, some delicate wild vegetables are burned to death in the dry and extremely hot weather with strong sunshine; in the last pentad, the wheat enters its mature period with heavy ears of wheat. With regards to agriculture, the Grain Buds is in the midsummer when the summer harvest, planting and management coincide and therefore, this period is also the busiest time for farmers in a year.

Customs

In the regions of Haining City in Zhejiang Province, there is a custom of water-capturing on the Grain Buds. Water-

水稻插秧
Rice Transplanting

时节，如果雨水不够丰沛，将会影响一年的收成。小满这天，各村都要举行抢水仪式，由村子里德高望重的人主持，召集各户黎明时分就燃起火把，在水车前放置香烛和各种面食，进行祭祀活动。祭祀完毕后，执事者以鼓锣为号，大家纷纷踏上事先安装好的水车，数十辆水车一起运转，将河水引灌入田。

蚕神是中国古代传说中司掌蚕桑的神，相传诞生于小满这天。中国最早开始种桑养蚕，这成为古代百姓的衣食之源。因此古时的人们把蚕视为天物，在小满时节举

capturing indicates a ceremony of irrigating the field, and praying for abundant rainfall and good harvest. In southern China, this is an important period for applying fertilizer to early season rice and transplanting semilate rice. Insufficient rainfall will impact the harvest of the whole year. On the Grain Buds, villagers will hold the water-capturing ceremony, and the highly respected person in the village is invited to host this activity. By gathering people up to light the torches and prepare candles and various cooked wheaten food before the waterwheels, all villagers conduct the sacrificial activities. After the ceremony, people start all waterwheels to irrigate the field based on the starting signal of the gong and drum given by the host.

The God of Silkworm is the god managing the business relating to silkworms in ancient legend. It is said that she was born on the Grain Buds. China is the country that began the mulberry tree planting and silkworm cultivation, regarded as the source of ancient Chinese people's food and clothes. Therefore, ancient people treated the silkworm as a thing from heaven and on the Grain Buds, the activity

● 蚕神庙
Temple of Silkworm God

行祭拜蚕神的活动，表达对蚕神的敬意，祈求养蚕有个好收成。小满日，江浙一带都要举行隆重的祭祀蚕神活动。在皇宫内，由皇后亲自主持祭祀；在民间，养蚕人家要到蚕神庙摆上酒菜、水果，祭拜蚕神像。

健康

小满节气值五月中旬，全国各地此时都陆续进入夏季，天气渐渐

of worshiping the God of Silkworm was prevailing to express their highest respect and pray for a wonder harvest in silkworm cultivation. On the Grain Buds, the area of Jiangsu and Zhejiang provinces would conduct colourful sacrificial activities. For the imperial palace, the worship ceremony would be hosted by the queen; and for the families who raised silkworms, they should provide offerings like liquor, dishes and fruits to the God of Silkworm in the temple.

炎热起来，雨水也开始增多。一方面人们贪凉，不注意防风、防寒，容易引发感冒发热咳嗽等病症，喜饮冷饮也容易引起肠胃方面的疾病；另一方面，炎热的天气使人体消耗增大，空气潮湿，容易导致人体内分泌紊乱，令风湿、肠胃疾病和皮肤病高发。因此在养生方面主要是防暑、防潮，注意健脾、养胃、利湿、清心、祛暑。在生活起居上，应顺应自然的变化，遵循夏季日长夜短的规律，保持充足的睡眠。还要改善居室环境，避免潮湿。而在饮食上应以健脾养胃为原

Health

The Less Fullness of Grain is during the middle of May when different regions in China step into summer successively, with increasingly hotter weather and more rainfall. On the one hand, it is easy to neglect the prevention from wind and coldness for more coolness and based on that, cold, fever and cough are likely to occur. What is more, the preference towards cold drinks is always the incentive for gastrointestinal diseases. On the other hand, hot weather causes an increase in the consumption of physical strength, and the high humidity

- 《梦蝶图》刘贯道（元）
Dreaming of the Butterfly, by Liu Guandao (Yuan Dynasty, 1206-1368)

则，以清淡、促食欲、易消化的膳食为主，多食用汤水和蔬菜水果。

can conduct the endocrine disorder, known as a high-incidence season of rheumatism, stomach disorders and skin diseases. Therefore, in terms of health preservation, the prevention of sunstroke and dampness, tonifying the spleen and stomach, inducing diuresis, clearing away the heat and driving away summer heat should be highlighted during this period. With regard to daily life, people should follow the natural law of longer daytime and shorter night and ensure sufficient sleep. Meanwhile, the indoor environment should be improved and dampness should be avoided. Lastly, according to the principle of tonifying the spleen and stomach, one should mainly choose food that is light and easy to be digested, and can promote the appetite, like soup, vegetables and fruits.

- **苦菜**

小满时节，正是苦菜收获的季节，吃苦菜也逐渐成为庆祝小满节气的饮食习俗。苦菜味道新鲜爽口，清淡香嫩，含有丰富的粗蛋白纤维和维生素C。另外，苦菜具有清热解毒、凉血、祛瘀止痛等功效。

Bitter Herbs

On the Less Fullness of Grain, the bitter herbs are ready to harvest. Eating the bitter herbs has gradually become the custom to celebrate the Solar Term. The bitter herbs taste refreshing and light, rich in crude protein fibre and Vitamin C. In addition, the bitter herbs can relieve internal heat and pain, cool the blood and remove stasis.

> 芒种

每年6月5、6或7日，太阳到达黄经75°时为芒种。"芒"顾名思义，指有芒作物的成熟，而"种"是指谷黍类作物的播种。农谚"有芒的麦子快收，有芒的稻子可种"，意思是说到了芒种，大麦、小麦等农作物已经成熟，迎来了收获的季节，而晚谷、黍、稷等夏播作物也迎来了播种最忙的时节。所以，芒种也称为"忙种"，是一年中农事最为繁忙之时。有俗语说："春争日，夏争时"，意思就是农民在这个时节要争分夺秒，忙着夏收、夏种和夏管。

我国古代将芒种节气分为三候：一候螳螂生，小螳螂在芒种时节破卵而出；二候䴗始鸣，喜阴的伯劳鸟开始在枝头鸣叫；三候

> Grain in Ear (*Mangzhong*)

Grain in Ear is on the 5th, 6th or 7th of June every year when the Sun travels to the 75° celestial longitude. *Mang* implies the maturity of awned crops and *Zhong* indicates the sowing of grain millet. According to the agricultural proverb, the awned crops like barley and wheat are ripe and can be reaped while the summer-sown crops like late-season cereal and millet are ready for the busiest season of sowing on the Grain in Ear. Therefore, this period is also known as the busiest time for sowing and agricultural activities. It is said that: every day counts in spring while every hour counts in summer, which means farmers should make full use of every minute to conduct summer harvest, sowing and management on this Solar Term.

The Grain in Ear consists of three

• 水田里劳作的农民
A Farmer Working in Paddy Field

反舌无声，反舌鸟与其他鸟类相反，在芒种这炎热的时节反而停止了鸣叫。

芒种时节已经正式进入夏季，除了炎热外，最显著的天气特征就是梅雨。到了芒种，南方就会进入阴雨绵绵的梅雨时节。因此时正是江南梅子的成熟期，故称为"梅雨"或"黄梅雨"。南宋诗人赵师秀在《约客》中就生动地描绘了梅雨时的景致："黄梅时节家家雨，

pentads: in the first pentad, the little mantis is born; in the second pentad, the shade-loving shrike sings on the tree; in the last pentad, the mockingbird stops singing in the burning summer, opposite to other birds.

From the Grain in Ear, summer officially comes. The most prominent features of the weather during this period include the scorching heat and plum rain. Southern China enters the plum rain season which is full of rainy days, when the plum in the south steps into the maturing stage. This is also the reason why the rain during this period is called plum rain or yellow plum rain. Zhao Shixiu, a poet from the Southern Song Dynasty (1127-1279), depicted the scenery of plum rain vividly in his poem *Inviting Guest*: During the plum rain season, it rains frequently. The croaking of frogs can be heard by the lawn or pond. My invited friend did not efface himself and midnight approached. I knock the chess piece while watching the falling of blossoms in the rain.

Customs

On the Grain in Ear, Chinese people are accustomed to offering sacrifices to the God of Flower as a tradition. The

青草池塘处处蛙。有约不来过夜半，闲敲棋子落灯花。"

习俗

芒种时节有送花神的习俗。农历二月十二是百花的生日，民间会举行迎花神的活动。到了芒种，已经是六月，百花过了最繁茂的花期，开始凋落，于是人们便在芒种这一天用花瓣枝条编成轿马，绫罗叠成伞状的旗帜，用彩线系在树枝

12th day of the second lunar month is the birthday of flowers and people will conduct various activities to worship the God of Flower. As the Grain in Ear is in June when the florescence is over and flowers begin to wither and fall, people employ branches and petals to produce the sedan chair, use silks to create umbrella-shaped flags and tie coloured strings on tree branches to worship the God of Flower on the Grain in Ear. Since many who attend the sacrificial ceremony are unmarried young daughters, this day

端午节

端午节是中国人纪念屈原的传统节日，为每年农历五月初五，也正值芒种节气。屈原是春秋战国时期楚国人，他忠君爱国，却遭受奸臣陷害排挤，不受楚怀王的信任，仕途不顺。公元前305年，由于反对楚怀王与秦国结盟，屈原被逐出楚国，开始流放生涯。公元前278年，秦军攻破楚国国都，屈原的政治理想破灭，便以死明志，在当年五月初五投汨罗江而亡。百姓们为了表达对屈原的敬意，将每年的农历五月初五定为端午节，这一天全国各地都举行盛大的活动，流传下来赛龙舟、吃粽子、喝雄黄酒、吃绿豆糕、煮梅子、悬挂菖蒲、蒿草、艾叶，熏苍术、白芷等习俗。

• 屈原蜡像
Wax Statue of Qu Yuan

- 粽子

粽子是端午节的传统节日食品，用粽叶包裹糯米蒸制而成。千百年来，吃粽子的习俗盛行不衰。每逢端午节，家家户户都要浸糯米、洗粽叶、包粽子。粽子的品种繁多，食材丰富，北方包糯米小枣粽，南方喜用鲜肉、火腿、蛋黄、豆沙等作粽子的馅料。

Zongzi (Chinese Rice Dumpling)

Zongzi is the traditional food on the Dragon-boat Festival, made out of steamed sticky rice wrapped in reed leaves. For thousands of years, eating rice dumpling has long been prevailing and popular among Chinese people. When the dragon-boat festival comes, every family will soak the sticky rice into water, wash reed leaves and wrap the rice dumplings. There are various kinds of rice dumplings with abundant materials. For instance, for northern China, people are accustomed to eat the rice dumplings with Chinese-date; for southern China, the rice dumplings with fresh pork, ham, yolk or sweetened bean paste are favoured.

Dragon Boat Festival

The Dragon Boat Festival is a traditional festival for Chinese people to commemorate Qu Yuan, on the 5th day of the fifth lunar month every year which is also the Grain in Ear. Qu Yuan was a poet of State Chu who lived during the Spring and Autumn Period and Warring States Period (770 B.C.-221 B.C.). He was patriotic and loyal to the throne but suffered from the calumniation and exclusion of treacherous officials. Based on that, he could not get the trust of King Huai of Chu whose official career was not smooth. In 305 B.C., since Qu Yuan opposed King Huai of Chu's opinion of forming the alliance with State Qin, he was expelled from State Chu and started his exile. In 278 B.C., the troops of State Qin broke through the capital city of State Chu, which disillusioned Qu Yuan's political ideal, and he had to express his thoughts through suicide. On the 5th day of the fifth lunar month in the same year, Qu Yuan drowned himself in the Miluo River and died. In order to deliver respect for Qu Yuan's spirits, people define the 5th day of the fifth lunar month as the Dragon Boat Festival and on that day, all regions across China will hold magnificent activities, like traditional dragon-boat racing, eating traditional Chinese rice-pudding (Zongzi, traditional Chinese rice dumpling) and green bean cake, drinking realgar wine, boiling plum, hanging calamus, wormwood leaves and folium artemisiae wormwood and incensing atractylodes and angelica root.

中国的花神

　　中国人认为花是吉祥之物，民间流传着许多与花神有关的习俗，如庆祝百花生日、迎送花神等等，各地还建有花神庙来祭祀花神。花神由农历月份确定，每月有一种时令花，称为"十二月花神"。各地流传的花神并不完全一致，较具代表性的说法是：一月兰花神、二月梅花神、三月桃花神、四月牡丹花神、五月芍药花神、六月石榴花神、七月荷花神、八月紫薇花神、九月桂花神、十月芙蓉花神、十一月菊花神、十二月水仙花神。

兰花（代表一月）
Orchid
(for the First Lunar Month)

梅花（代表二月）
Plum Blossom
(for the Second Lunar Month)

桃花（代表三月）
Peach Blossom
(for the Third Lunar Month)

牡丹（代表四月）
Peony
(for the Fourth Lunar Month)

芍药（代表五月）
Herbaceous Peony
(for the Fifth Lunar Month)

石榴花（代表六月）
Pomegranate Flower
(for the Sixth Lunar Month)

荷花（代表七月）
Lotus
(for the Seventh Lunar Month)

紫薇花（代表八月）
Crape Myrtle
(for the Eighth Lunar Month)

桂花（代表九月）
Sweet Osmanthus
(for the Ninth Lunar Month)

芙蓉花（代表十月）
Confederate Rose
(for the Tenth Lunar Month)

菊花（代表十一月）
Chrysanthemum
(for the Eleventh Lunar Month)

水仙花（代表十二月）
Narcissus
(for the Twelfth Lunar Month)

God of Flower in China

Chinese people generally believe that flowers are symbols of auspiciousness as a gift from Heaven. Therefore, there are many customs related to flowers, like celebrating the birthday of flowers and worshiping the God of Flower. Meanwhile, the temples of God of Flower are built in various regions in China. According to the lunar calendar, each month obtains its season flower, known as the Gods of Flower for Twelve Months. Although there is no unified opinion on the prototype of each God of Flower, the generally accepted practice is: God of Orchid for the first lunar month; God of Plum Blossom for the second lunar month; God of Peach Blossom for the third lunar month; God of Peony for the fourth lunar month; God of Herbaceous Peony for the fifth lunar month; God of Pomegranate Flower for the sixth lunar month; God of Lotus for the seventh lunar month; God of Crape Myrtle for the eighth lunar month; God of Sweet Osmanthus for the ninth lunar month; God of Confederate Rose for the tenth lunar month; God of Chrysanthemum for the eleventh lunar month and God of Narcissus for the twelfth lunar month.

上来祭奠花神。由于参加送花神仪式的多为闺中女儿家，因此芒种这一天也被称为"女儿节"。

在安徽南部地区，芒种时节还有安苗的习俗。这项习俗自明朝初期开始盛行并流传下来。到了芒种这一天，人们为了祈求新种的水稻能够丰收，就用新麦面捏成五谷六畜、瓜果蔬菜等形状，然后用蔬菜汁染上颜色，上笼蒸熟，作为祭祀供品，祈求五谷丰登、人畜平安。

健康

芒种时节天气炎热，湿度增大，人体消耗较大，容易疲惫、困倦，精神萎靡不振。因此要根据季节的特点，在生活起居上晚睡早起，注重午休，适当晒太阳，做运动，勤洗衣，多洗澡，以保持充沛的体能和良好的精神状态。在饮食方面，因夏季人体新陈代谢旺盛，汗易外泄，肠胃的消化功能也比较弱，因此要多吃能生津止渴、祛暑益气的食物。唐代名医孙思邈提倡："常宜轻清甜淡之物，大小麦曲，粳米为佳"，即应吃轻清甜淡的食物以保持体内外的清爽与平

is also known as the Daughter's Festival.

In the southern region of Anhui Province, there is a custom of seedling placement on the Grain in Ear. This custom has been prevailing since the beginning of the Ming Dynasty (1368-1644). People utilise fresh flour to create cooked wheaten food in different shapes like the five cereals, six domestic animals and various melons, fruits and vegetables, dyeing them into different colours. After that, people steam them and regard them as sacrificial offerings, praying for the harvest of newly planted rice seedlings and safeness for people and livestock.

Health

On the Grain in Ear, the weather is burning hot with high humidity and during this period, people suffer from huge energy consumption, weakness of limbs and drooping spirits, likely to feel tired. Therefore, according to the characteristics of the weather in this period, people should conduct a regular schedule as getting up and going to bed early, taking afternoon naps seriously, basking moderately, doing some physical exercises, frequent laundry and showering, in order to ensure the plentiful energy and wonderful spiritual

衡。炎热的夏天许多人缺乏食欲，因此摄取的食物应该营养丰富，易于消化，如山药、大米、大枣、鸡肉、牛肉、荞麦、豆类等。

statement. With regards to the diet, since people are likely to sweat due to the strong metabolism in summer, the digestive function of the stomach and intestine is relatively weak. Based on that, people should eat the food that helps produce saliva, slakes thirst, drives away summer heat and tonifies vital energy. Sun Simiao, a famous doctor in the Tang Dynasty (618-907), advocated that: it is suitable to choose fresh and sweet food, like barley and wheat, and polished round-grained rice is especially recommended. Through eating this kind of food, people can keep a healthy balance. In summer, the weather also leads to the loss of appetite and therefore, the diet should be rich in various nutrition and digestive food, like Chinese yam, rice, Chinese-date, chicken, beef, buckwheat and beans.

> 夏至

夏至是二十四节气中第十个节气，早在公元前7世纪，人们就用土圭测日影最先确定了夏至这一节气。

> Summer Solstice (*Xiazhi*)

Summer Solstice is the 10th Solar Term of a year. In as early as 7th Century B.C., people employed the earth sundial to calculate the length of the Sun shadows and determined the Summer Solstice as the first Solar Term.

The Summer Solstice is on the 21st or 22nd of June every year when the Sun travels to the Tropic of Cancer (23°26′ North Latitude), with the longest daytime in the Northern Hemisphere. *Zhi* indicates extreme and what the origin of *Xia Zhi* is was recorded in *Codex of Official Calendric System*: The Sun shines vertically on the extreme of the north in a year, with the longest daytime and shortest shadow. After the Summer

- 荷花盛开
  Lotus in Full Bloom

● 浙江杭州西湖苏堤日落时分
Sunset of West Lake Su Causeway of Hangzhou City in Zhejiang Province

每年的6月21日或22日，太阳直射地面的位置到达北回归线附近（北纬23°26′），北半球的白昼时间达到全年最长。所谓"至"，就是极的意思，《恪遵宪度抄本》对夏至名称的由来做了记载："日北至，日长之至，日影短至，故曰夏至。"意思是说夏至这一天，太阳直射地面位置达到一年的最北端，白昼时间最长，影子最短。等过了

Solstice, the direct solar radiation point moves towards the south and the daytime becomes increasingly shorter.

The Summer Solstice is divided into three pentads: in the first pentad, the antlers fall off; in the second pentad, the cicadas sing all day in the burning summer; in the last pentad, the shade-loving plants like pinellia ternate begin to grow prosperously and the heliophilous plants are declining. The weather will not become so hot before the arrival of

夏至以后，太阳直射位置将会向南移动，白昼时间逐渐变短。

夏至分为三候：一候鹿角解，到了夏至，鹿角开始脱落；二候蝉始鸣，炎热的夏天到来，蝉开始叫个不停；三候半夏生，一些同半夏一样喜阴的植物开始生长繁茂，而喜阳的植物开始衰败。"不过夏至不热"。到了夏至说明炎热的天气已经到来。此时气温高，日照、降水充足，是农作物快速生长的大好时机，同时也是害虫和杂草肆虐的时期，因此锄地成为农民的当务之急。

习俗

夏至是古代节日，有夏至祭地的习俗。到了这一天，古代帝王率

Summer Solstice, which indicates that the real burning days come with high temperatures and sufficient sunshine and rainfall, providing wonderful conditions for crops to develop. At the same time, pests and weeds are rampant and therefore, farmers are busy with hoeing during this period.

Customs

In the past, Summer Solstice was treated as a festival and celebrated with sacrifice activities. On this day, the Chinese emperor would take all civilian and

- **北京地坛方泽坛**
  地坛是明、清两代帝王夏至进行祭祀、祈求风调雨顺、国泰民安的地方。
  Fangze Temple at Temple of Earth in Beijing
  Chinese emperors in the Ming and Qing dynasties (1368-1911) held sacrifice ceremonies at Temple of Earth on the Summer Solstice to pray for good crop weather, peace and national prosperity.

● 夏至吃面是南北方共有的习俗
Eating Noodle on the Summer Solstice Day is a Custom Shared in both North and South China

领文武百官到郊外举行祭祀活动，以祈求灾消年丰。宋代还自夏至日起，百官放假三天。

自古以来，民间就有"冬至饺子夏至面"的俗语，有"吃过夏至面，一天短一线"的说法。夏至吃面在很多地区都非常盛行，这是因为到了这一时节，新麦入仓，所以夏至食面也有尝新的意思。北方多吃打卤面和炸酱面，而南方则吃阳春面、干汤面、肉丝面、三鲜面、过桥面及麻油凉拌面等。

另外，在浙江绍兴地区，夏至

military officials outside the capital city to hold a grand sacrifice ceremony in order to pray for a blessing as well as a good harvest. In the Song Dynasty (960-1279), officials would have three days off from the Summer Solstice day.

There is a folk custom called Dumplings in Winter Solstice and Noodles in Summer Solstice as well as a traditional saying that one finds that every day is shorter than the day before after having noodles on the Summer Solstice day. People from many places in China have noodle dishes on their Summer Solstice menu. This is because the harvest season of wheat ends before summer. Therefore having noodles made from newly harvested wheat on the Summer Solstice day symbolizes a taste of fresh delicacy. In northern China, people have thick wheat noodles topped with thick gravy or a mixture of ground pork and salty soybean paste; while in southern China, people have plain noodles, noodles in soup, noodles with shredded pork, seafood noodles, crossing the bridge noodles, cold noodles with sesame sauce, etc.

Additionally, people hold ancestral worship ceremonies on the Summer Solstice day in Shaoxing City, Zhejiang

有祭祖的习俗，称为"做夏至"。人们将新麦磨成粉，做成醮坨，用韭菜为佐料进行烹制，然后用竹签穿起来插放在田间的流水处，用来祭祀祖先。

健康

夏至时节，气候炎热，进入了一年当中阳气最旺的时期。人们受到气候影响，容易头晕恶心、胸闷心悸，缺乏食欲。因此在生活起居上，要晚睡早起，户外活动应该注意防暑防晒，避开正午烈阳。不要因为贪凉洗冷水澡或者过度吹风扇、空调，以免引起身体不适。在饮食方面，应以清淡为主，尽量少吃生冷食物，可适当吃些酸咸食物。

Province, which is called Summer Sacrifice. They make pastes with ground wheat and cooked them with Chinese chives. Those pastes are then placed along the water channels in the fields on bamboo sticks as sacrificial offerings.

Health

Summer Solstice marks the start of the hottest season in a year. One is easily affected by the hot weather and feels dizzy, nausea, chest tightness or losing appetite. Therefore, one needs to decrease the sleeping hours (go to bed late and rise early) and avoid prolonged exposure to high temperatures or doing physical activities under the strong sunshine at noon in case of heat stroke. On the other hand, taking cool baths, being overexposed to air conditioning or clinging to the fan for hours may also bring physical discomfort. In terms of the diet, light food is recommended and raw or cold food should be avoided, meanwhile the adequate intake of some sour or salty food is necessary too.

- 《松涧横琴图》朱德润(元)
Chinese Painting *Playing Guqin beside the Pine Trees and Stream*, By Zhu Derun (Yuan Dynasty, 1206-1368)

> 小暑

每年的7月6、7或8日，太阳运行位置到达黄经105°时为小暑。古代历书记载："斗指辛为小暑，斯

> Minor Heat (*Xiaoshu*)

Minor Heat begins on the 6th, 7th or 8th of July every year when the Sun reaches the celestial longitude of 105°. Ancient

云南哀牢山梯田
Terraced Fields in Mount Ailao in Yunnan Province

时天气已热，尚未达于极点，故名也。"意思是说小暑时天气已经炎热，但还没有达到最热，这是相对大暑而言。

小暑分为三候：一候温风至，到了小暑，风吹过大地，已经没有凉爽的感觉，而是带着热浪；二候蟋蟀居壁，由于天气炎热，蟋蟀开始躲在阴暗的角落避暑；三候鹰始鸷，天气越来越热，老鹰也开始在空中翱翔以降低自己的体温。

到了小暑，全中国进入炎热的季节，气温升高，雨水充沛，阳光充足，农作物茁壮成长，大部分地区的高粱、玉米、谷子等春种秋收的作物基本播种结束，农村多忙于夏秋作物的田间管理，因此又有"小暑开始热，减衣身上轻，抓紧种蔬菜，备足过严冬"的说法。

习俗

民谚有云："六月六晒龙衣，龙衣晒不干，连阴带晴四十五天。"在古代传说中，太阳要在六月初六这天为龙王晒龙衣，而农历六月六正值小暑时节。由于天气炎热潮湿，衣物容易发霉，人们

Chinese almanac describes when the handle of the Dipper points to *Xin* (the eighth of the ten Heavenly Stems, which was employed in the ancient Chinese astronomical positioning system to locate the celestial longitude of 105°), the temperature keeps going up and does not reach its peak, from which people name this solar term as Minor Heat. That is to say, the weather is hot during Minor Heat and begins yet not so hot as Major Heat.

Minor Heat has three pentads (each pentad lasts for five days). The Breeze turns warm in the first pentad: one finds that the wind becomes heat waves when Minor Heat comes. Crickets escape into crevices and cracks from the escalating heat in the second pentad. Eagles become more aggressive and are more often seen in the sky than ever in the third pentad, apparently to lower the body temperature against the escalating temperature when the third stage of Minor Heat arrives.

Lesser Heat leads the whole of China into the hottest time with rising temperatures, increasing rainfall and sunshine. The seedtime of crops harvested in autumn (sorghum, corn, millet, etc) in most areas in China has come to an end before summer and the consequent management of field crops

挂满晾晒衣物的上海弄堂（图片提供：微图）
People Living in Traditional Urban Alley Community in Shanghai Hanging Their Clothes Out

also needs time and labor. Hence there is a traditional saying that it starts to get hot from Minor Heat, and people grow vegetables to prepare for the winter in light clothes.

Customs

There is a traditional saying "The 6th day of the sixth lunar month is the day to hang the dragon robe out under the Sun and this day is followed by forty-five cloudy and sunny days". In the Han people's legendary stories, the God of the Sun will dry the Dragon King's robe, namely the rain god's robe on the 6th day of the sixth lunar month when Minor Heat begins. Due to the hot and damp weather, people move their clothes, quilts, containers, and scriptures out on sunny days in order to prevent the mould. There was once a tradition of bathing elephants in ancient China. In the Yuan Dynasty (1206-1368), the Ming Dynasty (1368-1644) and the Qing Dynasty (1616-1911), the emperor kept a team of honor guards made of chariots, horses and elephants.

便趁着这个时节天气好的时候晒衣服、棉被、器具和经文等。古时农历六月六还有晒象的习俗。元明清时期，皇帝拥有一支由车、马、象组成的仪仗队伍，每年的六月六这天，皇家侍卫用号旗和鼓乐将大象从宫中顺承门引导而出，进入河水之中，岸边还搭建彩棚，供仪官监督洗象的工作。

小暑正是小麦收仓、稻谷成熟收获的时节，为庆祝丰收，人们便用新收割的稻米做饭来祭祀五谷大神和祖先，然后人人吃尝新酒。又因"食新"在小暑节后的第一个天干为辛的日子（辛日），"新"与"辛"同音，因此称为"食新"，并流传着"小暑吃黍，大暑吃谷"的说法。

"伏"就是阴气被阳气压制潜伏的意思。三伏天是中国夏季气温

On the every 6th day of the sixth lunar month, the elephants were sent to the river for a bath through the Shuncheng Gate, with signal flags waving and music played at the gate. Then those elephants were bathed in the river with an etiquette official on the site in a colorful shed to supervise the operations.

Minor Heat is the time when people store wheat and harvest rice. In order to celebrate the harvest, people offer cooked rice (cooked with the new rice)

● 收割水稻的农民（图片提供：微图）
Peasants Harvesting Rice

最高、湿度最大的时节，分为初伏、中伏和末伏，通常初伏在小暑节气里。有句俗话叫做"头伏饺子二伏面，三伏烙饼摊鸡蛋"，在初伏开始的这一天，北方有吃饺子的习俗。在南方地区，有吃羊肉、吃炒面的饮食风俗。

健康

"小暑接大暑，热到无处躲"，小暑时节，气温较高，是人体阳气最旺盛的时候。人们在工作劳动时，要注意劳逸结合，尽量减少户

as a sacrifice to the God of Five Cereals as well as their ancestors and drink newly brewed rice wine, which is called *Shixin*, i.e. A Bite of Freshness (*Shi*, to eat or drink; *Xin*, freshness).The *Shixin* Day is the first day named after the Heavenly Stem *Xin* after Minor Heat. Since those two characters share the same pronunciation *Xin*, hence the name. There is also a traditional saying that one should eat millet on the Minor Heat day and wheat on the Major Heat day.

Fu refers to the period when the *Yang* energy suppresses the *Yin* energy, therefore the hottest and wettest summer days are named the triple-*Fu* days or dog days, divided into three stages: the first *Fu*, the second *Fu* and the third *Fu*. The first *Fu* arrives before Minor Heat ends. There is a traditional saying quoted as "one has dumplings in the first *Fu*, noodles in the second *Fu* and baked pancakes and scrambled eggs in the third *Fu*" in China. More specifically, in the first *Fu*, people from northern China have dumplings while people from southern China have mutton and fried noodles.

- 《槐荫消夏图》佚名（宋）
Chinese Painting: *Man Resting in the Cool Shadow of a Pagoda Tree*, by Anonymity (Song Dynasty, 960-1279)

夏季消暑茶
The Tea for Relieving Summer Heat

Health

There is a saying that: No one can escape from the heat, when Minor Heat and Major Heat arrive. After Minor Heat, the temperature is commonly high and it is the time when the *Yang* energy of the human body takes the power. When undertaking outdoor work, one should take a balance between work and rest; one's outdoor activities should not be less intensive in case of heatstroke. Hot weather also causes restlessness and fatigue, thus keeping a peaceful mind in both spring and summer is important, in which way the powerful *Yang* energy inside one's body remains under control. In addition, sultry and wet weather also increases the possibility of seasonal diseases including hemorrhoids, rheumatism and arthritis. It is therefore important to avoid catching a cold when cooling oneself. During this period, one needs to keep a light and digestive diet, having some heat-clearing food from time to time and less oily and spicy food. Heat-clearing soups and porridge are both good choices, which not only help with digestion but also stimulate one's appetite.

外活动，防止中暑。炎热的天气也容易使人感到烦躁不安，疲倦乏力，因此要坚持"春夏养阳"的原则，保持心态的平和。另外，此时天气闷热潮湿，容易引发痔疮、风湿和关节炎等疾病，因此在消暑的同时一定要避免受凉。饮食上以清淡、易消化为好，多吃祛暑食物，少吃油腻、辛辣的食物，各种消暑汤和粥是不错的选择，不仅有助于消化，同时也能够增加人的食欲。

> 大暑

大暑是二十四节气中最为炎热的节气。每年7月22、23或24日，太阳运行位置到达黄经120°时为大

> Major Heat (*Dashu*)

Major Heat, the hottest period among all twenty-four solar terms, begins on the 22nd, 23rd or 24th of July every year when the Sun is exactly at the celestial

• 绿树成荫
Trees Making a Pleasant Shade

- **郁郁葱葱的绿树**
  Green and Luxuriant Trees

longitude of 120°. According to the *Collection of Lunar Seventy-two Pentads* (each solar term can be divided into three pentads), *Shu* refers to heat, which is classified into minor and Major heat. July begins with minor heat and after the mid-July the heat grows greater; when the Major Heat begins, the temperature will reach its peak. Just as its name suggests, it comes with unbearable high temperatures.

Major Heat has three pentads. During the first pentad, baby fireflies break out of their eggs buried under decayed grass. During the second pentad weather turns sweltering and the land turns wet. The third pentad signified the storm season. Storms can slightly lessen the sultriness in the air and bring refreshing air. During Major Heat, the temperature, rainfall and sunlight all reach the annual peak. Thus this is the best time to grow thermophilic crops. Nevertheless, this is also the period when natural disasters (e.g. drought, flood and typhoon) happen most frequently. Fighting those natural disasters as well as field management therefore becomes the top priority.

暑。《月令七十二候集解》对大暑的解释为："暑，热也，就热之中分为大小，月初为小，月中为大，今则热气犹大也。"大暑，顾名思义，就是炎热至极。

大暑分为三候：一候腐草为萤，大暑时节，萤火虫破卵而出；二候土润溽暑，天气变得闷热，土地变得潮湿；三候大雨时行，时常有雷阵雨出现。每下一场暴雨，空气中的闷热就会稍微减弱。大暑时的气温、降雨和日照处于全年的最高峰，是喜温作物生长最快的时期，但旱、涝、台风等自然灾害也会频繁发生，抗旱、排涝、防台风和田间管理成为农忙中最为紧要的工作。

习俗

送大暑船是浙江台州一带的民间习俗。相传清代同治年间，大暑前后在台州一带经常病疫横行，人们怀疑是凶神作祟，于是就在江边

Customs

People from Taizhou City in Zhejiang Province will launch a boat on the river on the Major Heat day. According to legend, during the Tongzhi Period in the Qing Dynasty (1862-1874) there was an outbreak of epidemic disease in Taizhou City around the Major Heat. People at that time believed that the disease had been brought by evil spirits, so they built a temple where five gods were enshrined and launched a Major Heat boat hoping

● 送大暑船出海（图片提供：FOTOE）
Launching the Major Heat Boat

建了五圣庙，并送大暑船来祈求平安和渔业丰收。大暑船与普通渔船相似，大约8米长，2米宽，船内设置神龛和香案，供奉着祭品。大暑这一天，人们先举行迎圣活动，并在五圣庙做法事。随后，人们把大暑船运送到码头，举行祈福仪式，船起帆入海，然后在海上点燃，任其沉浮。

在大暑这一天，山东南部地区还流行着喝暑羊（即喝羊汤）的习俗。在古时候，大暑时节正好赶上麦收结束，在这个农闲的空隙，人们就呼朋唤友来喝羊汤，滋补身体。还有一种说法是，三伏天喝羊汤，喝得全身大汗淋漓，汗能带走人体内的积热，排除毒素，有益于健康。

而在福建地区，人们在大暑时吃荔枝、羊肉和米糟，称为"过大暑"。荔枝鲜嫩多汁，富含多种维生素和葡萄糖，具有较高的营养价值；而羊肉性温，对风寒咳嗽、体虚怕冷等病症有防治效果，是进补的佳品。米糟是中国特有的一种民间食物，由大米和酒曲发酵而成，性温甘辛，对调理脏腑有着较好的

● 羊肉汤
Mutton Soup

for well-being and a harvest of fish. The Major Heat boat is about eight meters long and two meters wide, similar to a fishing boat. There is a shrine and an incense table on which sacrificial offerings are placed. On the Major Heat day, people first have a ceremony to invite and welcome the gods with rituals going on at the five-god temple. Then they carry the boat to the dock and pray. In the end the boat is pushed out to the sea and put on fire.

On the Major Heat day, people in southern Shandong Province traditionally drink mutton soup. This is because in

• 米糟
Fermented Rice

效果。人们将红糖、母鸡等食材与米糟一起煮熟，用于进补。

健康

大暑时节，正值中伏前后，是一年温度最高、阳气最旺的时节。中国人历来有"冬病夏治"的说法，一些冬天容易发作的慢性疾病，如肺气肿、哮喘、风湿痹症等疾病，可以通过夏季的调养，使病情得到缓解，甚至痊愈。进入盛夏，天气炎热，人体消耗较大，抵抗力较低。在生活起居上应该劳逸结合，合理

ancient times, when people finished harvesting wheat around Great Heat, they took a rest and drank some mutton soup to get enough nourishment; there is another explanation for this tradition that one sweats heavily when drinking mutton soup on dog days, which would relieve heat and toxin from one's body.

In Fujian Province, people have lychees, mutton and fermented rice on the Major Heat day, which is called the Major Heat Diet. Fresh lychee is very nutritious as its juicy pulp is rich in vitamins and glucose. Mutton can be used as food therapy to tonify one's body and prevent wind chill, cough and bronchitis. Fermented rice is a unique Chinese folk cuisine. Cooked with rice and distiller's yeast, it helps to regulate visceral function. People sometimes cook fermented rice with brown sugar and hen for better nourishment.

Health

The hottest days in a year are those days around the Second *Fu* after the Major Heat day. It is believed among Chinese people that winter diseases can be cured with appropriate treatment in summer. That is to say, some seasonal chronic diseases mostly happen in winter

《莲塘纳凉图》金廷标（清）

Chinese Painting: *People Taking a Walk near a Lotus Pond*, by Jin Tingbiao (Qing Dynasty, 1616-1911)

安排作息，尽量避免长时间劳作，也不要贪凉，长时间待在温度过低的空调房间里。在饮食方面要注意补充水分，多喝白开水，少喝冷饮。食物要保持洁净，多吃豆类及蔬菜、水果等食物。

including emphysema, asthma and rheumatism which can be possibly relieved or even cured with treatment in summertime. In the midsummer when the temperature keeps climbing, the heat consumes one's energy fast and affects the weak immune system. Therefore it is important to keep a good balance between work and life as well as regular rest time. Long-time working under the Sun or hours spent in an air-conditioned room should be avoided in case of physical discomfort. About the diet, timely intake of water is vital. The boiled water is recommended instead of cold drinks. Food should be kept in clean condition. In terms of one's diet, it is better for one to take in more beans, vegetables and fruits in the summertime.

> 立秋

每年8月7、8或9日，太阳运行到黄经135°的日子为立秋。《月令七十二候集解》："秋，揪也，物于此而揪敛也"，意思是说到了立

> Beginning of Autumn (*Liqiu*)

Beginning of Autumn starts on the 7th, 8th or 9th of August every year when the Sun reaches the celestial longitude of 135°. According to the *Collection of Lunar Seventy-two Pentads*, Qiu（秋）

• 新疆喀纳斯的秋天
Autumn in Kanas in Xinjiang

新疆金秋白桦林
Birch Forest in Xinjiang in Autumn Time

秋，秋天正式开始，大部分作物已经过了最繁茂的生长期，将在立秋之后走向凋零。

立秋分为三候：一候凉风至，到了立秋，天气开始慢慢凉爽起来，风也渐渐散去暑气，带来一丝清凉；二候白露生，早晚的温差开始变大，早晨空气中的水分凝结成雾气笼罩着大地；三候寒蝉鸣，喜阴的寒蝉开始鸣叫。

refers to shrinking and holding back, which means that Beginning of Autumn signifies the start of autumn and most crops stop growing and start to wither.

Beginning of Autumn is divided into three pentads. During the first pentad, it turns cool and the wind takes away the residual heat. During the second pentad, the temperature gap expands between the day and night, which causes water vapor to condense into the mist on autumn mornings. When the third pentad comes, people hear winter cicada singing.

When Greater Heat ends and Beginning of Autumn arrives, the temperature does not drop as it should. Strangely it will remain hot for one to two weeks although the autumn has already started, which is called the Tiger Autumn. This short-term lingering heat and strong sunshine is followed by real autumn weather, i.e. cool and mild days. In the past, people call this period of time (from Beginning of Autumn to the Autumn Equinox) the Long Summer. Crops grown in the spring and summer are now harvested, and autumn-harvested crops grown in the autumn are thriving, which also needs careful attention. It is the busiest season for field workers. Agricultural idiom such as "one sunny

立秋之后的天气比较特别。俗话说"暑去凉来",但实际上立秋时节天气并没有快速凉爽下来,反而会短期回热。天气晴朗、阳光强烈的暑热天气会持续一两周,人们称这种天气为"秋老虎"。"秋老虎"之后,天气从夏天向秋天过渡,逐渐变得温和舒适。古时候人们将立秋到秋分这段时间称为"长

day at Beginning of Autumn saves a lot energy of field workers", "three rainfalls at Beginning of Autumn can turn blighted grain into plump grain", and "rainfalls at Beginning of Autumn bring gold all over the field" all vividly illustrate the importance and decisive role of good weather in harvest time.

Customs

Holding ceremonies to welcome the arrival of autumn is a traditional Chinese ritual, which can date back to the Zhou Dynasty (1046 B.C.-221 B.C.). Ancient Chinese employed the west out of the five orientations (east, west, south, north, middle) and the color white out of the five colors (cyan, red, white, black, yellow) to represent autumn. At the Beginning of Autumn, the emperor led his civil and military officials out of the capital city to the west and held a grand ceremony to celebrate the arrival of autumn, worshiping the White Lord and Autumn God. In the Han Dynasty (206 B.C.-220 A.D.), this ritual was imbued with other meanings: when the ceremony was taking place, people killed livestock as sacrificial offerings and the emperor would hold a hunting game as a display of his absolute military power. Later the

• 《耕织图册・祭神》焦秉贞(清)

Illustrated Book of Farming and Weaving: God Worshiping, by Jiao Bingzhen (Qing Dynasty, 1616-1911)

夏", 春夏耕种的作物迎来了收获的季节, 大秋作物进入重要的生长发育阶段, 农事进入最为繁忙的季节。"立秋晴一日, 农夫不用力""立秋三场雨, 秕稻变成米""立秋雨淋淋, 遍地是黄金", 这些农谚都生动地说明立秋时节的天气对农作物收成起到决定性的作用。

习俗

迎秋是古老的节令活动, 早在周代就有迎秋的习俗。古人将秋和五方之中的西方、五色之中的白色相对应。立秋这天, 天子率领文武百官去西郊举行隆重的迎秋仪式, 祭祀白帝和秋神。到了汉代, 迎秋被赋予了新的含义, 在仪式上人们要杀牲畜来做祭品, 天子要射猎, 表示秋来扬武之意。后来, 立秋风俗广泛传播, 活动内容也多样化起来。立秋时, 人们吃一些新鲜的瓜果, 以消暑气。

秋社是秋天祭祀土地神的习俗活动。立秋正是秋收时节, 百姓为了庆祝丰收, 会举行盛大的仪式来酬谢土地神。秋社自汉代就开始流

folk customs at the Beginning of Autumn were widely accepted and fully developed by the public, for instance, people have some fresh fruits on this day to drive away residual heat in their bodies.

Autumn sacrifice is a traditional activity to worship the God of Land in the autumn. The Beginning of Autumn is the time for harvest and people will hold grand ceremonies in celebration of harvest as well as in gratefulness of the God of Land. Autumn sacrifice can date back early to the Han Dynasty (206 B.C.-220 A.D.). Later in the Song Dynasty (960-1279), in addition to sacrifice activities, people would also have cakes and wines, and the newly married bride would visit her parents at the Beginning of Autumn. The famous poet Lu You in the Song Dynasty has given a lively description of autumn sacrifice in his poem *Sacrificed Livestock* as follows: on the sacrifice day, people kill and cook pigs as sacrificial offerings in my village. The air is filled with a pleasant smell, which attracts hungry birds to gather on the tree branches and the temple gate. We do not have big ceremonies but we follow the tradition of asking villagers to take some meat back home and share the blessing with their children.

行，到了宋代，秋社除了祭祀土地神之外还增加了食糕、饮酒、妇女归宁等风俗。宋代诗人陆游的《社肉》："社日取社猪，燔炙香满村。饥鸦集街树，老巫立庙门。虽无牲牢盛，古礼亦略存。醉归怀余肉，沾遗遍诸孙"，就生动描述了秋社的热闹场景。

健康

立秋之后，天气开始变得凉爽干燥，但白天仍炎热潮湿，还是应该注意防暑和除湿，以免引病上身。在生活起居上，可以随着季节的规律早睡早起。立秋后早晚温差变大，天气忽冷忽热，应遵循"春捂秋冻"的原则，根据天气变化适度增减衣服。在饮食方面，切忌盲目贴秋膘、大量进补，这样会加重肠胃负担，导致消化功能紊乱。可以食用一些温润滋补、强健脾胃、清肺的食物，如银耳、蜂蜜、芝麻等。此外，秋天是万物肃杀的季节，此时人的情感比较丰富敏感，所以要调养好精神，保持乐观开朗的心态。

Health

After the Beginning of Autumn day, it turns cooler and drier while the weather remains hot and humid during the day. Therefore heatstroke prevention and dehumidification are still necessary. It is better to keep early hours in the autumn. During this period of time, the temperature changes sharply on different days and the temperature gap between day and night is great, one should decide on daily clothes based on the old Chinese saying Warm in Spring and Cool in Autumn, which means that one should not decide what to wear on hearing what season it is, or ignore the daily changes of temperature, which could be very sudden and sharp in the early spring or early autumn. In terms of one's diet, one should not have too big meals that may increase the burden on one's digestive system and cause digestive disorders. It is suggested that one take in some mild food, food that protects and strengthens one's spleen and stomach and food that clears away the lung heat, such as white fungus, sesame and honey. Additionally, one may easily become sentimental when seeing the bleak scene in autumn. Thus keeping an open and optimistic mind is very important for one's mental health.

七夕节

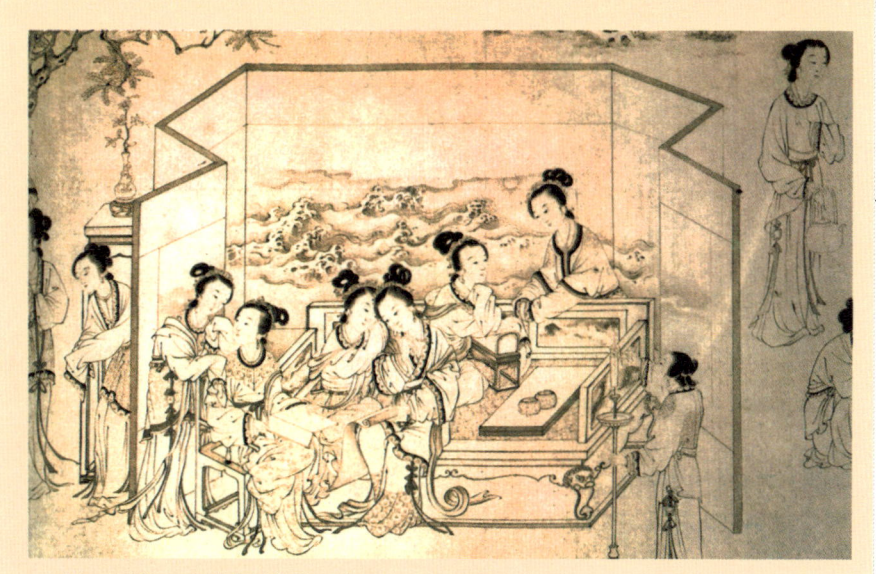

　　七夕节又称"乞巧节""女儿节",是在农历七月初七,正值立秋时节。七夕节历史悠久,起源于汉代。相传七夕是天上牛郎和织女相会的日子。牛郎与织女的故事是中国流传千古的爱情故事。织女是天上的仙女,一次下凡与牛郎结识并相爱,二人结为夫妻,生下一双儿女。不料被织女的母亲王母娘娘得知,强行把织女带回天宫。牛郎追上天界,却被王母娘娘用银簪划出的银河拦住去路。牛郎与织女隔河相望,天各一方。喜鹊被他们的爱情感动,化作"鹊桥"让牛郎与织女团聚。王母娘娘也为之动容,允许二人在每年农历七月初七在鹊桥上相会。每年七夕节,民间女子在庭院中摆上桌案,供奉香烛、瓜果、女红等,向织女祈求智慧和巧艺以及浪漫真挚的爱情。可以说,七夕节是中国具有浪漫色彩的节日。七夕节经久不衰,到了宋元时期,在当时的国都设有乞巧市,从七月初一就开始贩卖各种乞巧物品,集市上车水马龙,热闹非凡。

• 《乞巧图》丁观鹏【局部】(清)
Chinese Painting: *Talent Praying*, by Ding Guanpeng [Part] (Qing Dynasty, 1616-1911)

Double Seventh Festival (Chinese Valentine's Day)

The Double Seventh Festival, also called the Talent Praying Festival or the Girls' Festival falls on the seventh day of the seventh lunar month during the Beginning of Autumn period. This festival was firstly celebrated in the Han Dynasty (206 B.C.-220 B.C.). The seventh day of the seventh lunar month is believed as the once in a year opportunity for two legendary lovers, cowherd and weaver girl, to meet each other. Their romantic story has a thousand-year history in China: the weaver girl was a fairy girl who had escaped from the heaven and later fell in love with cowherd, a cowherd. They had a happy marriage and two children. However, the weaver girl's mother, the Queen Mother of the West, was furious when knowing this marriage and she forced the weaver girl to go back with her. The husband followed them to the heaven but was unfortunately stopped by the Milky Way, which was transformed from the Goddess's hairpin. However, their love story moved the a flock of magpies, who decided to help them and flied up to form a bridge over the Milky Way on which the couple could be together. Then the Queen pitied on them and she agreed that the couple could spend one single night every year with each other, i.e. the seventh day of the seventh lunar month. When celebrating the Double Seventh Festival, girls placed candles, fruits and their needlework pieces on the altars in their courtyards, hoping for wisdom, sewing skills, and true love. The Double Seventh Festival is considered as a romantic festival celebrated by Chinese people. In the Song Dynasty (960-1279) and Yuan Dynasty (1206-1368), capital cities opened Talent Praying Market, where people could buy necessary objects to prepare the celebration. This market opened on the first day of the seventh lunar month, crowded and busy all the time.

二十四节气
Twenty-four Solar Terms

- 《秋林舒啸图》颜峄（清）
  Chinese Painting: *Chill Wind of Autumn Whistling in the Forest*, by Yan Yi (Qing Dynasty, 1616-1911)

> 处暑

处暑是二十四节气中反映气温变化的节气。每年的8月22、23或24日，太阳运行到达黄经150°时为处暑。《月令七十二候集解》

> End of Heat (*Chushu*)

End of Heat, whose name reflects changes in daily temperature during this solar term, starts on the 22nd, 23rd or 24th of August every year when the Sun

• 云南元阳梯田
Terraced Fields in Yuanyang in Yunnan Province

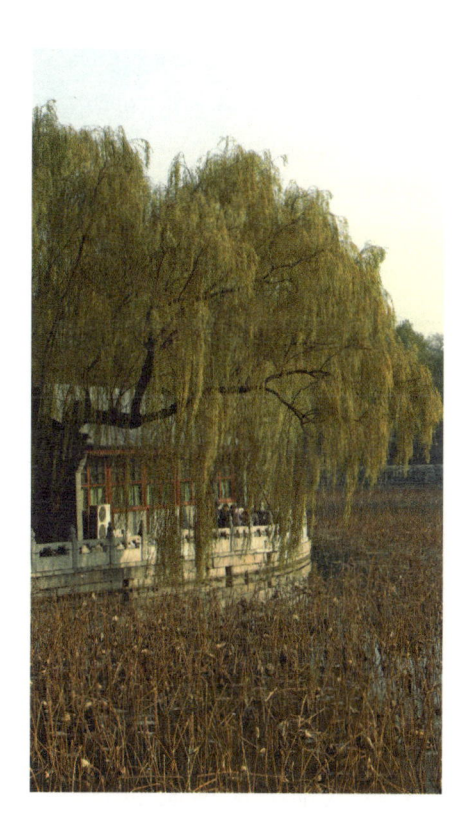

北京北海残荷
Withered Lotuses in Beihai Park in Beijing

对处暑的解释是："处，止也，暑气至此而止矣。"处是结束的意思，处暑顾名思义就是夏天结束。在处暑时节，由于太阳对地面的直射位置持续向南移动，因此天气逐渐由炎热向寒冷过渡，夏天正式结束了。

处暑分为三候：一候鹰乃祭鸟，天气变凉，老鹰开始大量捕捉鸟类为过冬做准备；二候天地始

reaches the celestial longitude of 150°. It is recorded in the *Collection of Lunar Seventy-two Pentads* that *Chu* refers to the end of summer heat. Therefore, End of Heat signifies the end of summer. When End of Heat arrives, the location of direct solar radiation keeps moving southward which causes temperature drops, claiming the end of summer.

The End of Heat is divided into three pentads. When the first pentad comes, the temperature begins to drop and eagles hunt birds to prepare for the coming winter. During the second pentad, most plants stop growing and start to wither. The third pentad unveils the harvest season: after a whole year's hard work, field workers are now busy collecting their fruitful results. From End of Heat, evenings become cooler while days remain warm, which helps the crops to ripen, just as the saying goes: crops only ripen with the help of

肃，大地上的作物经过春夏生长繁茂期开始步入凋零期；三候禾乃登，处暑是收获的季节，经过一年的辛勤耕耘，农民迎来了丰收时节。处暑之后，暑气消退，昼暖夜凉的气候条件对于庄稼来说也是快速成熟的大好时期。农谚有"秋不凉，籽不黄"，在这秋高气爽的金秋时节，"处暑禾田连夜变"，田间作物成熟得很快，农民对农作物的管理进入了最后阶段。

习俗

在处暑前后的农历七月十五为中元节，是三大鬼节之一。发展至今，已成为人们祭奠祖先的节日，民间的习俗是放河灯。在中元节的夜晚，人们三五成群将亲手扎制的纸灯放入河道，用来悼念逝去的亲人。

处暑时节正是渔业丰收的旺季。自古以来就有开捕祭海的习俗。每到这个时候，在东部沿海地区就会举行盛大的开渔节，不仅有庄严肃穆的祭海仪式，欢送蓄势待发的渔民出海捕鱼，还开展各种文

chilly autumn. This season is considered the gold time for crops, which ripen very quickly. A traditional saying describes that "with only one night the crops may ripen during End of Heat". It also means that the field work now enters the final stage.

Customs

Around End of Heat on the 15th day of the seventh lunar month, people celebrate the Dead Spirit Festival, one of the three major ghost festivals in China. It has now become a festival for people to worship their ancestors by floating candlelit lanterns on the river. When the night falls, people gather together and put their hand-made paper lanterns on the river along with their miss and love for the deceased family members.

End of Heat is also the peak season for the fishing industry. Historically and presently people hold sacrifice activities for the blessing of the god of the ocean. Every year around this time, people from the eastern coastal area of China will celebrate the Fishing Festival as well as other cultural, tourism and trade activities, which has attracted many tourists and business people.

- 《渔家乐》年画
  New Year Painting: *Happy Fisherman*

化、旅游和经贸活动，吸引无数游人、商客前往。

健康

处暑正值金秋时节，天气由炎热向凉爽过渡，空气从湿润转向干燥，人们也要随着气候的变化调整生活节奏。俗话说"春困秋乏"，在秋季人体的新陈代谢开始变慢，开始储存能量，因此在生活起居上应该保证充足的睡眠时间，早睡早起，适当的午睡可以缓解秋乏。此时户外秋高气爽、气候宜人，做些登山、散步、打太极拳之类的运动

Health

End of Heat falls in the autumn when the weather turns cool and dry, therefore the way of health preserving should also adapt to seasonal changes. That one feels fatigued in spring and autumn is a traditional Chinese saying, which reflects the fact that the metabolism of the human body begins to slow down in the autumn for storing energy. One should keep adequate and regular rest time, and sometimes taking a nap also releases seasonal fatigue. During this period, the weather is ideal for outdoor activities including hiking, walking and playing Taijiquan (a traditional Chinese shadow

● 公园中太极拳练习者
People Practicing Taijiquan at Park

对身体有益。秋季气温降低，早晚温差变大，但要遵循"春捂秋冻"的原则，不宜急于添加衣服。在饮食方面，应吃些清淡、有营养、滋阴润燥的食物。

boxing) as the air is clear and refreshing. As the temperature drops, there is an expanding temperature gap between day and night. However, one should also consider the possible daily changes in temperature when deciding what to wear. In terms of daily diet, one should take a light and nutritious diet and have some heat-clearing food.

> 白露

每年的9月7日前后，太阳运行到达黄经165°时为白露。古代历书上记载："斗指癸为白露，阴气渐重，凌而为露，故名白露。"意思是到了白露时节，天气转凉，清晨空气里的水分都凝结成了白色的露珠，因此称为"白露"。《月令七十二候集解》记载："水土湿气凝而为露，秋属金，金色白，白者露之色，而气始寒也。"可见白露是天气转凉的象征。

白露分为三候：一候鸿雁来，白露之后天气明显转凉，按照气候变化迁徙的大雁成群结伴飞往南方准备过冬，因此也有"八月雁门开，雁儿脚下带霜来"的说法；二候玄鸟归，燕子从北方飞回到南方；三候群鸟养羞，鸟儿开始积极

> White Dew (*Bailu*)

White Dew begins from around the 7th of September every year when the Sun reaches the celestial longitude of 165°. It is recorded in the ancient Chinese almanac that when the handle of the Dipper points to *Gui*, people see drops of dew formed by condensed water vapor and therefore people name this solar term as White Dew. To be more specific, White Dew was named because during this period of time, the temperature drops so that water vapor in the air condenses into drops of dew in the early morning. According to the *Explanation of the Seventy-two Pentad*, "atmosphere's moisture coming from groundwater and soil condenses into droplets of water, which is called dew. Autumn belongs to the element Metal, which is represented by the color white. White in turn is the

• 辽宁秋收时节
Harvest Season in Liaoning Province

觅食，储存食物为过冬做准备。有俗语说："处暑十八盆，白露勿露身。"意思是处暑时节还可以用盆洗浴，但是到了白露时节，就不能赤膊了。白露之后，太阳直射地面位置向南移动，日照的强度和时间都逐渐减少，天气已经转凉。但白露也是一年当中最好的时令之一。古代诗歌总集《诗经》上说"蒹葭苍苍，白露为霜"，为人们描绘出一幅恬静而辽阔的秋季景象。各处

color of dew. When people see dew, the temperature drops." This is why White Dew signifies the temperature-dropping period.

White Dew is divided into three pentads. The wild geese come when the first pentad arrives: when the White Dew falls, people see the southward migration of wild geese, just like the traditional saying that the eighth lunar month launches the migration of wild geese with dew on their feet. The big birds fly south

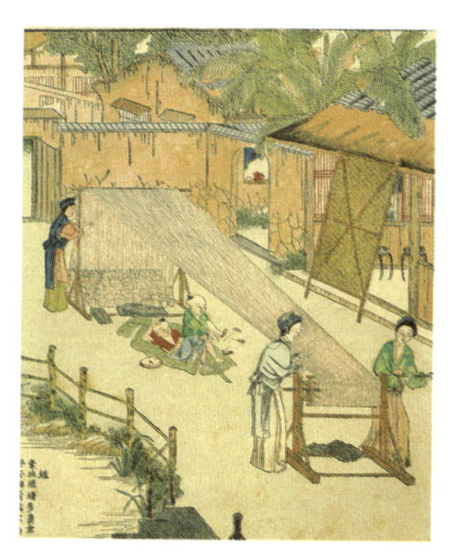

• 《耕织图册·经图》焦秉贞（清）
Illustrated Book of Farming and Weaving, Wrapping Threads, by Jiao Bingzhen (Qing Dynasty, 1616-1911)

• 《耕织图册·织图》焦秉贞（清）
Illustrated Book of Farming and Weaving: Weaving, by Jiao Bingzhen (Qing Dynasty, 1616-1911)

when the second pentad comes: during the second pentad swallows fly from northern China to southern China. Birds stock their hoards when the third pentad begins: at this time of year birds actively forage for food and hoard food for winter use. There is a folk saying that during End of Heat, one bathes oneself with a basin every day, while eighteen days later when White Dew starts it is impossible to do so due to the chilly weather. At this time of the year, the direct solar radiation keeps moving southward, which leads to a reduction of sunlight intensity and time. Although the weather turns cold, White Dew is still believed as the best time of the year. In the *Classic of Poetry*, a sentence reads, "in the blossoming dark reed there are some droplets of dew", which depicts the quiet and grand picturesque scenery in the early autumn. The clear and refreshing weather and beautiful mountains and rivers all invite people to go on an autumn trip. After a year of hard work, field workers are happy to collect their abundant agricultural products including sorghum, cotton, and corn which ignite the harvest season and beautifully decorate the vast fields. At the same time, the plantation of winter wheat is also about to begin.

都秋高气爽，山明水秀，是秋游的大好时节。而经过一年的辛勤劳作，大地上一派丰收景象：火红的高粱、雪白的棉花、黄澄澄的玉米……将田间点缀得如锦似画。在迎来收获的同时，冬小麦的播种也即将开始。

习俗

江苏太湖地区有白露时节祭祀禹王（又称"大禹"，是中国古代传说中距今5000多年前的部落联盟领袖）的习俗。禹王是传说中的治

Customs

During White Dew, people from the Taihu Lake area in Jiangsu Province hold sacrifice activities in memory of Yu the Great (also called Great Yu, a legendary ruler in ancient China around five thousand years ago). He had spent great effort dredging the riverbeds, successfully controlled the flood-hit areas along the Yellow River, Yangtze River and Huai River and trapped the devil turtle in the Taihu Lake. Because of his contribution, Yu the Great has been worshiped as a great hero. To express their gratitude

● 太湖风光
Landscape of Taihu Lake

浙江绍兴大禹陵
Mausoleum of Yu the Great in Shaoxing City, Zhejiang Province

水英雄，他疏通三江，有效地治理了从黄河到江淮的水患，传说他将兴风作浪的鳖鱼镇压在太湖底下。百姓为了表达对禹王的感激之情，每到白露日就举行盛大的活动来祭祀他，祈求平安和丰收。附近渔民从四面八方赶到位于太湖中央小岛的禹王庙举行祭祀活动，后来逐渐发展成为热闹的庙会。

"补露"是在白露时节注意滋补身体的意思。各地进补的食物多种多样。在江苏南京，人们喜欢喝白露茶。茶树经过夏季的酷热，到了白露前后正是生长的最佳时期。白露茶不像春茶鲜嫩但不禁泡，也不像夏茶干涩味道苦，而有着独

and pray for peace and harvest, people living in the Taihu Lake area hold great ceremonies on the day of White Dew. Their sacrifice activities once took place at the Temple of Great Yu, which is located on an island in the center of Taihu Lake and later these activities were replaced by temple fairs.

Traditionally, one should better keep a nourishing diet when White Dew comes, which is called Nutritious Dew. People from different places follow different diets: in Nanjing City, Jiangsu Province, people drink White Dew tea. The best growth season for tea trees is after the hot summer ends, i.e. White Dew. Unlike a teabag of spring tea that is fresh yet can only make one cup of good tea or the summer tea that tastes dry and bitter, the White Dew tea smells fragrant and tastes sweet; in many places in Jiangsu and Zhejiang provinces, people make White Dew wine with grains such as glutinous rice and sorghum to share with their visitors; and in Fujian province,

● 采制白露茶
Picking Tea Leaves to Make White Dew Tea

特甘醇的清香味道。江苏、浙江地区一到白露，家家户户用糯米、高粱等五谷杂粮酿酒，俗称"白露酒"，用来招待客人。福建地区有在白露日吃龙眼的习俗，民间认为龙眼有益气补脾、养血安神、润肤美容的功效。

健康

白露时节，天气变得凉爽，在生活起居方面要注意天气变化，早晚适当增加衣物，让身体尽快适应气温的变化，提高耐寒能力。体质较弱的儿童和老人以及关节炎、支

people eat longans on the White Dew day as they believe that eating longan strengthens one's spleen, soothes one's nerves and smoothes one's skin.

Health

When the weather turns cold, one should add some clothes and prepare for dropping temperature. Those whose health conditions are not very good especially children, senior citizens and patients with arthritis, bronchitis and cardiovascular or cerebral vascular diseases need to keep warm. One should intensify outdoor activities and improve one's immune system. The dry weather

气管炎、心脑血管疾病的患者要注意保暖。白露时节总体上气候宜人，应适时加强体育锻炼，提高身体免疫力。由于气候干燥，人们容易咽干、咳嗽、皮肤干燥，而天气变凉也易导致过敏性疾病发作，因此在饮食上要多吃润肺、养肾、富含维生素的蔬菜水果，如梨、苹果、葡萄、龙眼、柿子等。过敏性体质的人，应少吃或者不吃海鲜、腌菜等生冷咸肥的食物。

may cause dry throat, cough and dry skin and when the weather turns cold, the possibility of allergy attacks increases. In case of those seasonal discomforts, one should have more vegetables and fruits to clear one's lung heat, strengthen kidney function and supplement vitamins, such as pear, apple, grape, longan and persimmon. People with allergies should better keep their hands off raw, cold, salty or fatty food, such as seafood or pickles.

• 云南大理白族老人
Aged Bai People in Dali City, Yunnan Province

> 秋分

每年的9月22、23或24日前后，太阳运行到达黄经180°的日子为秋分。汉代董仲舒所著的《春秋繁露》中记载："秋分者，阴阳相伴也，故

> Autumn Equinox (*Qiufen*)

Autumn Equinox begins on the 22nd, 23rd or 24th of September every year when the Sun reaches the celestial longitude of 180°. According to the *Spring and Autumn Annals* written

• 黑龙江北大荒
Great Northern Wilderness in Heilongjiang Province

• 辽宁本溪的秋天
Autumn in Benxi City, Liaoning Province

by Dong Zhongshu in the Han Dynasty (206 B.C.-220 A.D.), "the Autumn Equinox achieves the balance between *Yin* force and *Yang* force, therefore day and night are equally long and cold and heat are equally gentle." It has told two facts: firstly, the Autumn Equinox claims the end of the first half of autumn; secondly, on this day the Sun shines directly on the equator, which means that the length of day and night is equal. After this day, the direct solar radiation point continues moving southward and days are shorter in the northern hemisphere with day and night temperature gap increasing and temperature dropping. The late autumn is on its way.

Autumn Equinox has three pentads. In the first pentad, thunder begins to soften: thunderstorms are less frequent when Autumn Equinox arrives. In the second pentad, insects make nests: as the weather gets cold, insects make nests underground where they lie dormant or keep eggs during the winter. In the third pentad, water begins to dry up:

昼夜均而寒暑平。"一是秋分日将秋季平均分为两部分；二是在秋分这天太阳直射赤道，昼夜平分。秋分之后，太阳直射地面的位置将继续南移，北半球的昼短夜长现象越来越显著，昼夜温差逐渐变大，气温逐日下降，步入深秋季节。

秋分分为三候：一候雷始收声，秋分之后，打雷的现象减少；二候蛰虫坯户，天气转冷，虫类开始在地底下的洞穴里蛰伏，准备产卵过冬；三候水始涸，天气越来越干燥，江河湖海里的水量明显变少，一些较浅的水洼甚至干涸了。进入秋分时节，"一场秋雨一场寒"，北方有些地区已经见霜，大部分地区晴空万里、风和日丽，而西南地区则迎来了阴雨连绵的多雨季节。在这样丰富多变的天气条件下，农民依然是最忙碌的。汉末崔寔在《四民月令》中写道："凡种大小麦，得白露节可种薄田，秋分种中田，后十日种美田。"农民不仅要忙于秋收，还要忙于秋耕和秋种。北方忙于种冬小麦，南方则忙于种水稻。农民要积极"抢早"，为防止霜冻和绵雨灾害，要抢收秋收作物，而冬作物则是越早种越有利于生长，为来年的丰收打下基础。

习俗

秋分是古代传统的"祭月节"，到了这天，历代帝王都要举行祭月活动。后来达官贵族、文人墨客也相继仿效，渐渐传入民间。每年秋分时节，人们对着天上的月亮观赏祭拜，咏物抒情。

秋分还有送秋牛的习俗。所谓秋牛，就是一张印着农历节气

the dry weather takes away water from the ground and some shallow puddles may even dry up. There is an old saying that a rain of autumn is a rain of chill. After Autumn Equinox, morning frost will appear in some areas of Northern China but in most places in the North, people can enjoy the bright sky and good weather. By comparison, southwestern China enters the rainy season. Despite the fickle weather, peasants are still the busiest. According to *National Calendric System* written by Cui Shi in the late Han Dynasty, "in terms of the plantation of wheat and barley, one cultivates barren fields around White Dew, secondary fields around the Autumn Equinox and best fields during the last ten days of autumn." One can therefore tell that peasants are not only busy with harvest but also autumn ploughing and seeding. In the north, people plant winter wheat, and in the south, people plant rice. Timing is vital: the earlier they finish, the less their crops will be affected by frost and endless rainfalls. When hurrying to harvest crops, they also have to finish winter crop plantation as soon as possible, so as to seek the best harvest in the next year.

- 北京月坛鼓楼

月坛是明、清两代皇帝祭祀夜明神（月亮）和天上诸星神的场所。

Drum Tower at the Temple of the Moon in Beijing

The Temple of Moon was employed by emperors in the Ming and Qing dynasties (1368-1911) to worship the God of Night (the Moon) and the Gods of Star.

和耕牛图案的红纸，称为"秋牛图"。在秋分这天，能言善唱的人挨家挨户送秋牛图，每到一家就见机行事，说一些应景的吉祥话以讨取赏钱，这种活动也称为"说秋"，是人们庆祝和祈求丰收的习俗。

Customs

On the Autumn Equinox day, all emperors in ancient China held moon ceremonies with sacrificial offerings in all dynasties. Officials, noblemen as well as scholars also followed and celebrated this day, and later this ritual was gradually accepted by the public. Traditionally, Autumn Equinox is celebrated by Chinese people in many ways including sacrifice activities and poem-making under the glorious full moon.

Another tradition on the Autumn Equinox day is called Giving Autumn Cattle. The Autumn Cattle, also called the picture of Autumn Cattle, is a piece of red paper drawn with patterns of the twenty-four solar terms and a cattle. On the Autumn Equinox day, street performers hand out the pictures of Autumn Cattle from door to door. When the door opens, they pass up pictures and give an auspicious speech to earn some tips. This is also called the Autumn Speech and is considered a way to celebrate and pray for harvest.

中秋节

每年农历八月十五是中国的中秋佳节。"中秋"最早记载于《周礼》，意为秋天的中旬。古代将每个季节分为孟、仲、季三部分，中秋也称为"仲秋"。中秋节源于古代祭月的礼制，古代帝王"春分祭日、夏至祭地、秋分祭月、冬至祭天"。但是秋分在农历八月里的日子每年不同，不一定都有圆月，因农历八月十五的月亮较其他月份的更圆、更皎洁，所以祭月的仪式就从秋分调到八月十五。

到了唐代，中秋节作为节日固定下来。在唐代，每到中秋佳节，人们都要祭拜月神，赏月叙谈。宋朝时期，每逢中秋节，人们都要穿上盛装，赏月、品食月饼。明清时，祭月的仪式更加隆重，人们准备好月饼、葡萄、西瓜等祭品，焚香燃烛，祈求家庭和睦，团团圆圆。有的地方还有玩花灯、烧斗香、树中秋、点塔灯、舞火龙等风俗。经过几千年的累积，中国人的中秋情节非常浓厚。每当八月十五，看着天上朗朗明月，人们祈盼合家团圆，身在异乡的游子更是渴望与家人团聚，思乡之情溢于言表。

Mid-Autumn Festival

The Mid-Autumn Festival is held on the fifteenth day of the eighth lunar month every year. The earliest record of mid-Autumn is in the *Rites of Zhou*, in which it was interpreted as the middle period of autumn. Ancient Chinese divided every season into three stages, namely *Meng* (the first), *Zhong* (仲, the second) and *Ji* (the third). Therefore the two *Zhong* (中 and 仲) characters are both used in *Zhongqiu*. The Mid-Autumn Festival stemmed from the ancient Chinese ritual ceremony of the Moon. In the past, the emperor held the sun's ceremony on the Spring Equinox day, earth's ceremony

- 月饼寓意团圆美满
  Mooncake Implies Family Unity and Happiness

on the Summer Solstice day, moon's ceremony on the Autumn Equinox day and heaven's ceremony on the Winter Solstice day. However the Autumn Equinox day falls on different days every year and does not have the full moon every time, therefore the 15th day of the eighth lunar month was chosen to hold moon's ceremony. It was believed that this day had the brightest full moon.

The Mid-Autumn Festival has become an official festival from the Tang Dynasty (618-907), when people had sacrificial offerings to the God of the Moon and got together and had a good talk under the Moon. In the Song Dynasty (907-960), people would get dressed and eat mooncakes on the Mid-Autumn Day. In the Ming and Qing dynasties (1368-1911), people celebrated this festival in a more elaborate way: they would prepare mooncakes, grapes and watermelons as sacrificial offerings and light up incense and candles, praying for family peace and happiness. In some places, there are more customs including lantern festival, burning tower-shaped incense, lifting bamboo sticks with paper lanterns, lighting up lanterns on the pagoda, dragon parade, etc. Chinese people share a very strong cultural tie, which is particularly obvious in the celebration of Mid-Autumn Festival. On the 15th day of the eighth lunar month, family members get together and spend this precious moment together. Those who spend that day in the foreign land are also missed or greeted, as their love for their families and homelands is always remembered and cannot be described with words only.

- 《月曼清游图之琼台赏月》陈枚（清）
  Chinese Painting: *Evening Tour: Enjoying the Moon on the Heavenly Platform*, by Chen Mei (Qing Dynasty, 1616-1911)

健康

秋分过后,中国大部分地区真正进入秋季,日照减少,气温降低,日夜平衡,在生活起居上要注意保暖,同时坚持锻炼身体,增强体质,提高免疫力。秋天人们的肠胃功能比较弱,一方面天气转凉容易使肠胃炎症发作;另一方面,丰富的食物和过度进补也容易使食物不易消化。所以在饮食上,应该多吃一些时令瓜果蔬菜以及芝麻、核桃、糯米等养阴润燥的食物。尽量少吃辛辣、油腻的刺激性食物。

Health

Most places in China enter the autumn after the Autumn Equinox day, which means reducing sunlight exposure and declining temperature. It is beneficial to keep warm and do more exercises in daily life. The function of one's intestine and stomach may be affected, as the weather gets cold and one's plate is filled with harvested agricultural products. Therefore, one should take in more seasonal vegetables and fruits to release the burdens on the digestive system. At the same time, one ought to have more heat-clearing food such as sesame, walnut and glutinous rice and less spicy or oily food.

- 晾晒玉米的妇女
  Woman Hanging Corns

> 寒露

　　每年10月8日或9日太阳运行到达黄经195°的日子为寒露。"寒"就是寒冷，"露"就是露水，古代通常用"露"来表达天气转凉变冷之意。所谓寒露，意思是说天气变冷，地面上的露水快要凝结成霜了，有"寒露寒露，遍地冷露"的俗语。

　　寒露分为三候：一候鸿雁来宾，寒露时节，北方的气候开始变得寒冷，候鸟已经迁徙到南方准备过冬；二候雀入大水为蛤，雀鸟都销声匿迹，海边却出现了与雀鸟的颜色条纹相似的蛤蜊，民间相传这是雀鸟变的；三候菊有黄华，寒露时节迎来了菊花的花期。《月令七十二候集解》中记载："九月节，露气寒冷，将凝结也。"到了寒

> Cold Dew (*Hanlu*)

Cold Dew starts on the 8th and 9th of October every year when the Sun arrives at the celestial longitude of 195°. *Han* means cold, and *Lu* means dew and the latter was used by ancient Chinese to refer to the start of the cold season. Therefore, Cold Dew comes from the fact that at this time of the year, the weather turns further cold and droplets of dew on the ground form into frost. Just as the traditional saying goes, Cold Dew comes with frozen dew.

　　Cold Dew has three pentads. The first pentad brings guest geese to visit: during this period of time wild geese have finished their migration to south China. The second pentad sends sparrows to the ocean where these birds become clams: people could not find sparrows and when they saw clams with similar colors on the beach, they believed that

• 云南东川红土地
Red Earth in Dongchuan District in Yunnan Province

露，天气明显转凉，尤其是北方，从凉爽到寒冷的过程短暂而迅速。东北和西北地区已经提前进入冬季，部分地区甚至开始出现银装素裹的雪景。而南方仍旧风和日丽，金桂飘香，雨水减少让天气格外晴朗宜人，与北方形成了鲜明的对比。寒露时节的气候变幻莫测，特有的秋绵雨、寒露风和高原雪灾会使一年的辛勤耕作功亏一篑，因此在晴朗的天气抢收抢种、防灾减害成为农事主要的内容。

the disappearing sparrows had turned into clams. The third pentad starts the flower season of chrysanthemums. According to *Collection of Lunar Seventy-two Pentads*, in September, dew will be frozen due to the low temperature. One can tell the changes in the weather when the Cold Dew starts, particularly in northern China. It takes only a few days to get quite freezing. Northeastern and northwestern China has already entered into winter and some places are even covered by snow, which is greatly different from the weather in

习俗

寒露前后的农历九月初九为重阳节。重阳节历史悠久,早在战国时期就已经形成。百姓在重阳节这天有登高望远、插茱萸、赏菊饮酒的习俗。

由于"久久"和"九九"谐音,重阳节也是祭祖和敬老的佳节。重阳节最具代表性的食物是重阳糕和菊花酒。重阳糕又称"花糕""五色糕",用糙米、枣面等夹杂核桃、果脯等各种干果制成。人们还把菊花与发酵的粮食混合在一起,酿制成菊花酒,以待来年的重阳节饮用。这些习俗延续至今,依然兴盛不衰。

● 茱萸
Cornel (*Zhuyu*)

southern China: warm and sunny, and the air is filled with a sweet fragrance of Osmanthus. Getting less rainy, the weather in the south is beautiful and mild. It is not easy to tell the next day's weather in Cold Dew, which creates difficulties in fieldwork. As continuous autumn rain, Cold Dew wind and snowstorms in plateau areas may have a devastating influence on ripened agricultural products, it has become the top priority to finish harvesting ripened products as well as seeding winter crops as soon as possible with effective disaster prevention measures.

Customs

The Double Ninth Festival on the ninth day of the ninth lunar month was celebrated early in the Warring States Period (475 B.C.-221 B.C.). Traditionally, on this day people will go hiking, wear cornel, appreciate chrysanthemums and drink homemade wine.

As eternity and double ninth share the same pronunciation *Jiujiu* in Chinese, the Double Ninth Festival is also an opportunity to worship the ancestors and care for the elderly. People make a special menu for this day, i.e. Double Ninth cake

健康

寒露时节，天气明显转凉，气温下降极快，降雨减少，天气变得干燥。在生活起居上，天气变冷，人的睡眠时间增加。古人道："秋三月，早卧早起，与鸡俱兴。"就是要适当调节睡眠时间，早睡早起，有益于身体健康。此时，人的抵抗力下降，不宜进行剧烈的运动，在适度的体育锻炼后要注意保

and chrysanthemum wine. The Double Ninth cake, also called Flower Cake or Five-color Cake, is made with half-polished rice, date flour as the main body and decorated with dried fruits such as walnuts and preserved fruits. People also make wine with chrysanthemums and fermented grains, which is stored for one whole year. When the next Double Ninth Festival falls the wine will be ready to serve. This tradition is still followed by many Chinese now.

Health

During the Cold Dew season, the temperature goes down quickly. It becomes less rainy and dry, which affects the human body greatly. When it turns cold, one feels like getting more sleep. Ancient Chinese believed that in autumn, one should keep early hours like a rooster for the sake of better health. Adjusting one's rest time and keeping early hours does help enhance one's health. The immune system of the human

- 《月曼清游图之重阳赏菊》陈枚（清）
Chinese Painting: *Evening Tour: Chrysanthemums Appreciation on the Double Ninth Day*, by Chen Mei (Qing Dynasty, 1616-1911)

- 《补沈周重阳酒兴诗图轴》钱榖（明）

  Scroll of Chinese Painting and Poems about Our Gathering on Double Ninth to Shen Zhou, by Qian Gu (Ming Dynasty, 1368-1644)

body is also affected by the weather, and thus it will be wise to undertake moderate exercise rather than intensive exercise during this period of time. After doing physical activities, one needs to put on more clothes to keep warm and avoid catching cold. There is a folk saying that "during White Dew one needs to keep the body warm and during Cold Dew one needs to keep the feet warm". That is to say, the blood supply of feet is weak where heat is quickly consumed. Once the heat of feet is not well preserved, one may get weak or even sick. This is why it is very important to keep one's feet warm on cold days, i.e. Cold Dew. Keeping a balanced and healthy diet also benefits one's health: nutritious and dryness-clearing food such as walnut, sesame, white fungus, radish, lotus root and lily are all good choices; at the same time, chicken, duck, beef, fish, and shrimp also strengthen one's body.

暖，以免受凉感冒。民谚有"白露身不露，寒露脚不露"之说。寒从足生，人的双脚离心脏远，血液供应弱，而且脚部一旦受凉会使人体抵抗力下降，所以寒露时节要特别注意脚部保暖。在饮食调养方面，以核桃、芝麻、银耳、萝卜、莲藕、百合等滋阴润燥的食物为佳，注意增加鸡、鸭、牛肉、鱼、虾等食物以增强体质。

- 润燥食材
  Food for Moistening Dryness

> 霜降

霜降是秋天的最后一个节气。每年的10月23日前后太阳运行到达黄经210°的日子为霜降。《月令

> Frost's Descent (*Shuangjiang*)

Frost's Descent is the last solar term of autumn. It starts on about 23rd of October when the Sun arrives at the celestial longitude of 210°. Just like the *Collection of Lunar Seventy-two Pentads* describes: in the mid-ninth lunar month, the water vapor condenses into dew and the dew forms the frost. That is to say, the ninth lunar month is rather chilly and people wake up to see heavy frost.

The Frost's Descent has three pentads. With the arrival of the first pentad, jackals and wolves spend all day hunting to prepare for the coming winter. With the coming of the second pentad people see withering grass and fallen

- 红叶
Red Autumn Leaves

七十二候集解》中写道："九月中，气肃而凝，露结为霜矣。"到了农历九月，天气变冷，空气中的水分已经凝结成霜。

霜降分为三候：一候豺乃祭兽，天气渐渐凉了，豺狼开始大量捕获猎物，囤积脂肪为过冬做准备；二候草木黄落，草木经过夏天的繁华，郁郁葱葱的枝叶开始衰败，枯萎落地；三候蛰虫咸俯，虫类在这个时节一反夏季的嘈杂热闹，蛰伏在地底下，用细土将洞穴封起来等待来年春天的到来。到了霜降节气，在黄河流域会出现第一场霜，而南方离初霜还有一个月左右的时间。此时北方大部分地区已经秋收扫尾，而南方正是大忙时节，收获早稻、晚稻，栽种冬麦、油菜，采摘棉花，翻耕整地。

习俗

在我国南方很多地区有霜降吃柿子的习俗。"霜降吃丁柿，寒冬不流涕"，民间认为霜降吃柿子能够预防感冒。霜降是食用柿子的最佳时节，柿子在霜降前后完全成熟，此时的柿子皮薄、肉多、汁

leaves: in summer all plants thrive and now they fall into decay. With the start of the third pentad, all insects go into hibernation: dormant in the soil, insects seal the entrances of their underground caves with a layer of soil and wait for the next spring in the caves. Frost's Descent brings the first autumn frost of the Yellow River basin. One month later, the southern China will have its first autumn frost. Field workers in most places in the north have almost finished harvesting ripen crops, while their counterparts in the south have just begun their busiest days: harvesting early rice and late rice, seeding winter wheat and rape, picking cotton and ploughing fields.

Customs

In southern China, most people eat persimmon after Frost's Descent day. A folk saying goes like this: a persimmon in Frost's Descent keeps the cold away. This saying comes from people's belief that eating persimmon can prevent them from catching cold in autumn. Around Frost's Descent, persimmon is completely ripened and the ripe fruit is fleshy and pulpy with sweet and fresh juice. Its rich sugar and vitamins

甜，新鲜可口。霜降正值深秋，人体容易秋燥，而柿子富含糖分和维生素，可以起到润肺生津、清热止血的功效。

柿子美味，但含有鞣质，切勿与海带、紫菜、酸菜、黑枣、鹅肉、螃蟹、甘薯、鸡蛋、萝卜等食物一起食用，避免引起身体不适。另外，吃柿子的时候不能饮酒，以免中毒。

help clear lung heat and relieve cough, especially in late autumn when one's respiratory system is easily affected by dry weather.

Persimmon is delicious. However, due to the tannins in its pulp, it cannot be served with the following food: kelp, seaweed, pickled cabbage, black date, goose meat, crab, sweet potato, egg and radish. In addition, eating persimmon with alcohol will cause food poisoning.

- 在晾晒的柿子串中玩耍的儿童（图片提供：FOTOE）
Children Playing Next to Strings of Persimmons

健康

　　霜降正是季节变换的时节，气温变化剧烈，昼夜温差增大，受寒凉天气影响，人体皮肤和血管收缩，促使血压升高，容易诱发心绞痛、中风等疾病。所以在生活起居上应注意保暖，适当添加衣物。还应注意体育锻炼，以增强体质，可以选择登山、散步、打拳等户外运动，既锻炼了身体，又呼吸了新鲜空气。在饮食方面，由于天气干燥，应注

Health

Frost's Descent is a solar term between two seasons, i.e. autumn and winter when temperature changes violently and the temperature difference between day and night increases. This may cause skin shrinkage, vasoconstriction and hypertension and increase the possibility of angina pectoris and stroke. Therefore one should wear thick clothes to keep warm. Physical activities also improve one's immune system. One can go hiking, go walking or do Taijiquan to strengthen one's health and at the same time get

虎戏
Tiger Exercises

鹿戏
Deer Exercises

熊戏
Bear Exercises

意补充水分，食用梨、苹果、柿子、百合、山药、藕、荸荠等生津润燥、清肺化痰的食物，少吃辛辣、凉性食物。霜降是进补的好时节，民谚"一年补透透，不如补霜降"，可以适当食用鸡、鸭、牛、羊肉。

some fresh air. In terms of daily diet, one should take adequate water, some dryness and lung-heat cleaning food as well as sputum-reducing food, such as pear, apple, lily, yam, lotus root and water chestnut. At the same time, one should better eat less spicy and cold food. Frost's Descent is also a good season to have some nourishing food such as chicken, duck, beef and mutton. A traditional Chinese saying describes: nourishing food taken in during Frost's Descent supports a whole year.

猿戏
Monkey Exercises

鹤戏
Crane Exercises

• 五禽戏

五禽戏是东汉名医华佗发明的仿生健身术，通过模仿虎、鹿、熊、猿、鹤的动作，疏通人体经络和气血，以达到强身健体的目的，对改善神经系统和心肺功能有益。

Exercise of the Five Animals

The Exercise of the Five Animals is a health-preserving way invented by Hua Tuo, a famous doctor in the Eastern Han Dynasty (25-220) from studying movements of the tiger, deer, bear, ape and crane. It works by strengthening the function of collateral channels of human body and tonifying one's *Qi* and blood. One's neural system and cardio-pulmonary function will be improved after practicing this exercise.

> 立冬

立冬是冬天的第一个节气，与立春、立夏、立秋一样是代表季节开始的节气。每年11月7日或8日，太阳运行到达黄经225°，为立冬。

> Beginning of Winter (*Lidong*)

Beginning of Winter, the first solar term of winter, signifies the start of a season like Beginning of Spring, Beginning of Summer and Beginning of Autumn. It starts on the 7th or 8th of November every year when the Sun reaches the celestial longitude of 225°.

It is divided into three pentads. In the first pentad, water begins to freeze: with temperature falling below zero, water freezes naturally. In the second pentad, land freezes too. In the third pentad pheasants enter the ocean and become clams: in the past when the third pentad started, people could not see big birds such as pheasants, and when they found

● 黑龙江乌苏里江日出
Sunrise on the Ussuri River in Heilongjiang Province

• 长江风光
Landscape of Yangtze River

立冬分为三候：一候水始冰，气温降到零度以下，水面开始结冰；二候地始冻，土地也开始结冻；三候雉入大水为蜃，雉即野鸡之类，蜃为大蛤。立冬之后，野鸡一类的大鸟便不多见了，在海边可以看到外壳与野鸡的线条及颜色相似的大蛤，古人认为是雉变成了大蛤。

《月令七十二候集解》对"冬"的解释是："冬，终也，万物收藏也。"意思是说从立冬开

clams on the beach with similar color and stripes to pheasants, they believed that those clams were transformed from the disappearing pheasants.

Collection of Lunar Seventy-two Pentads explains that winter is the end of the year when all things rest. It means that since Beginning of Winter, all plants have withered under the bitterly cold weather; all harvested crops have already been collected and dried to keep in barns; animals hide in warm places for a winter-long sleep. The Beginning of Winter

始，万物凋敝，秋季作物全部收晒完毕，收藏入库。动物经过一年的生长繁衍之后进入了漫长的休整期，躲藏起来准备冬眠。立冬代表漫长的冬季即将开始。由于中国地域辽阔，各地的天气有着明显的差异，东北已经进入寒冬季节，大雪封山，土地封冻；长江中部则刚刚进入晚秋时节，天气开始慢慢变冷；华南地区迎来一年中最为清爽宜人的时节，阳光灿烂，雨水减少。因此，当东北的农作物进入越冬时期的时候，中部地区的农民正在进行"三秋"（秋收、秋耕、秋种）的收尾工作，华南地区的农民则开始抢种冬麦，清沟排水，准备越冬防冻等工作。

习俗

自古以来立冬就是人们非常重视的节气。古人认为这天如果天气晴朗，那么这个冬天就会非常寒冷；如果天气阴霾，就会迎来一个暖冬。自周朝起，就有在立冬这天于北郊举行盛大的庆典活动来祭祀黑帝的习惯。因为古人将冬季与北方、玄色相配，而黑帝又是代

claims the coming of winter, whose pace differs in different parts of China: in the northeast, mountains and plains are covered by heavy snow; while places along the middle reaches of the Yangtze River have just been visited by late autumn and dropping temperature; at the same time, the southern China has its best weather of a year with great sunshine and reducing rainfall. As a result, when the wintering season of crops is undergoing in the north, field workers in central China are still busy finishing their three autumn tasks, i.e. harvesting, ploughing and seeding whilst people in southern China have just started seeding winter wheat, dredging irrigation canals and ditches and preparing to prevent possible frost damage.

Customs

Since antiquity, the Beginning of Winter has been a very important solar term to Chinese people. Ancient Chinese would practice divination on this day: the Sun on this day was believed to bring a cold winter and clouds were believed to bring a warm winter. Since the Zhou Dynasty (1046 B.C.-221 B.C.), a grand ceremony to worship the Black Emperor would be held at the Beginning of Winter. This is

立冬进补（图片提供：微图）
People Having Meals at the Beginning of Winter

北方的五帝之一。典礼上，皇帝还会赏赐大臣冬衣以及其他的体恤恩典。

人们为了增强体质、抵御寒冬，习惯在立冬这天进补，俗称"补冬"。各地补冬的食物因地域的气候特点和风俗习惯而有所不同。北方人在立冬这天吃倭瓜馅的饺子，民间认为瓜代表结实，吃倭瓜馅饺子，可以滋补身体。而南方

because in the past, people believed that winter was represented by the direction north and the color black. As a result, the Black Emperor, one of the five northern Great Emperors would be worshiped at the Beginning of Winter. When the ceremony was taking place, the emperor would hand out winter clothes and other necessities to his officials.

To get physically prepared for the coming winter, Chinese people will have

人烹制各种鸡鸭鱼肉，姜母鸭、炖羊肉等冬季进补食物很受欢迎。

健康

古人对"冬"的解释是"万物收藏"，因此在立冬时节，也讲究休养生息、养精蓄锐。在生活起居上，应该保持规律的作息习惯，早睡晚起，保证充足睡眠和充沛的精力。多穿衣物，尤其注意背部和关节的保暖。冬季是进补的好时节，

some nourishing food at the Beginning of Winter, which is called the winter supplement. Based on the local weather and customs, different places have different winter supplement dishes: in the north, people eat pumpkin dumplings. It is believed that pumpkin represents strength and eating pumpkin dumplings will enhance one's health; in the south, most people cook meat dishes. Duck stewed with wine and ginger as well as stewed mutton are both popular dishes for winter supplements.

• 冬季进补也要讲究荤素搭配 (图片提供：图虫创意)
Balanced Diet for Winter Supplement

中医认为冬季进补能促进新陈代谢，提高人体的免疫力，改善身体畏寒、畏冷的现象。但冬天进补应遵循一定的原则，不能随意乱补，要顺应自然，在饮食上可以多吃温热养阳、热量高、营养丰富的食物，如牛肉、羊肉、鸡肉等，提高身体的御寒能力。另外，可以同时吃一些新鲜的瓜果蔬菜，以补充维生素和纤维素。

Health

In ancient times, Chinese people interpreted winter as all things in the state of preservation. Therefore, one should also follow the basic concept of rest to conserve strength and store energy by getting regular rest time and adequate sleep. As for the clothes, keeping one's back and joints warm is vital. Traditional Chinese medical science emphasizes winter supplements which can promote metabolism, improve the immune system and relieve chilly sensations. However, winter supplements should abide by the basic principle of balance. Warm food, calorie-rich and nutritious food such as beef, mutton and chicken can help the human's body to resist the cold air in winter. Furthermore, fresh fruits and vegetables with rich vitamins and cellulose are also wise choices in wintertime.

> 小雪

　　小雪是二十四节气中反映天气现象的节气。每年11月22或23日，太阳运行到达黄经240°的日子为

> Minor Snow (*Xiaoxue*)

Minor Snow describes the weather condition during this solar term. It begins on the 22nd or 23rd of November every

- 黑龙江的冬天
  Winter in Heilongjiang Province

小雪。古籍《群芳谱》中写道："小雪气寒而将雪矣,地寒未甚而雪未大也。"《月令七十二候集解》记载："十月中,雨下而为寒气所薄,故凝而为雪。小者未盛之辞。"意思是说小雪时节,天气变得寒冷,降水形式由雨转为雪,但降雪量还不大,由于地面气温不够低,下的雪是半冰半融状态的,落到地面就即刻融化了。

小雪分为三候:一候虹藏不见,小雪时节天气已经颇为寒冷,北方以降雪为主,南方虽然下雨,但是雨后的彩虹已经销声匿迹了;二候天气上升地气下降,天气寒冷导致天空的阳气上升,地面的阴气下沉,阴阳两气不能融会贯通;三候闭塞而成冬,小雪时节的北方已经封冻。这时由于地面温度还不够低,小雪的来临通常是冰粒夹杂着雨滴,俗称"雨夹雪",雪在地面的停留格外短暂。而南方刚刚进入初冬,雨水明显减少。农谚说:"小雪雪满天,来年必丰年。"意思是说,如果小雪节气下雪的话,预示着来年的雨水会比较均匀,适合作物生长。而且积雪一方面可以

year when the Sun comes to the celestial longitude of 240°. It is recorded in *Book of Flowers* that "in Minor Snow the weather gets bitterly cold and people see light snow falling, but not heavy snow"; and the *Collection of Lunar Seventy-two Pentads* describes that "in the middle of the tenth lunar month, the chilly air freezes raindrops so it snows, yet not heavily." Both historical records reflect that during Minor Snow, raindrops freeze into snow particles and quickly melt before they hit the ground as it is not cold enough yet.

Minor Snow has three pentads. During the first pentad, rainbows disappear from view: when winter starts, most places in northern China see snowy days and southern China see rainy days without rainbows. During the second pentad, the *Qi* of the sky ascends and the *Qi* of the earth descends: low temperature causes the *Yang* energy of the sky to go up and *Yin* energy of the earth to go down. Hence the two become clearly separated and cannot consummate with each other. During the third pentad, closure, namely the end of mixing between *Yin* and *Yang*, leads to the winter: Minor Snow freezes northern China, but the ground temperature is not

杀虫减害，另一方面也起到了保温的作用，有利于肥沃土地。这个时期，大部分的收割、播种已经结束了，主要的工作是防冻，修剪作物的枝干，包裹草秸。农闲时节还可晾晒蔬菜，进行草编、造肥等农副业工作。

习俗

在中国南方地区，有小雪时节吃糍粑的习俗。糍粑是一种传统食

low enough to freeze the falling snow particles. Therefore the snow will melt into raindrops before hitting the ground, which is called snowy rain or rainy snow. An agricultural saying tells that snow in the Minor Snow season suggests next year's harvest, which means that if it snows in Minor Snow, the rainfall of the next twelve months will be ideally equal and people will have a harvest year. The snow cover kills pests, keeps the covered crops warm and maintains field fertility. Most fieldwork has finished during this time of the year and the major tasks have become frost prevention, trimming crops and wrapping grass stalks. In Minor Snow, people also process harvested vegetables, make straw braids or produce fertilizers in their spare time.

Customs

In the south, people have glutinous rice cakes during Minor Snow. This type of cake is a traditional food, firstly used as sacrifice and now as festival food. Before cooking, people soak the rice for a while. After it cooks, the rice is

- 打糍粑（图片提供：FOTOE）
Pounding Glutinous Rice Paste

物，最早用来祭祀，现在变成了庆祝节日的食物。人们将糯米放在清水中浸泡，然后煮熟再放到石臼或者石槽里用木杵舂，直到捣成泥状。各地制作糍粑都有自己的特色，有的油炸，有的添加馅料。

小雪时节，土家族有杀年猪、吃刨汤的传统风俗，有庆祝丰收的意思。每年腊月，人们把自家的猪杀了，取一块最好的肉，和许多杂菜，炖一大锅，请左邻右舍、亲朋好友吃一顿，俗称"吃刨汤"。

健康

小雪时节虽然天气寒冷，人的"内火"却是最为旺盛的时候。人们用层层冬装将自己裹得严严实实，又长期待在温暖的室内，加上寒冷的天气给大快朵颐提供了最好的理由。因此，许多人在这个时候反而最容易上火。小雪时节在生活起居上，要规律作息、适度锻炼，多寻找兴趣爱好，培养积极乐观的生活态度。在饮食方面，适当吃一些养肾食物，如党参、黄芪、鹿茸、桂圆、腰果、芡实、山药等。

pounded into a rice paste in the stone scoop (or sink) with a wooden pestle. Different places have different ways to make glutinous rice cake: some may fry the cakes and some may add stuffing in the cakes.

During the Minor Snow season, the Tujia people traditionally kill pigs and have some pork soup as a harvest celebration: in the twelfth lunar month, people kills a pig, selects the best meat to stew soup with vegetables and shares the soup with neighbors, relatives and friends.

Health

Despite the cold weather, the internal heat of the human body reaches its the peak during Minor Snow. People put on heavy winter clothes and spend all day in their warm houses. Due to the bitter cold, one has every reason to have some nourishing food and this is why one also easily suffers from excessive internal heat. Regular rest time, adequate exercise, pleasant habits and a positive attitude all helpful for keeping healthy. Kidney nourishing food is best for health in winter, such as hairy asiabell root, astragalus root, pilose antler,

- **冲泡普洱茶**

  普洱茶的茶性温和，具有降脂、养胃、消食、养颜等功效，非常适合冬季饮用。

  Making Pu'er Tea

  Pu'er tea is a mild tea and best to drink in winter months. It suppresses blood fat and invigorates one's stomach as well as the digestion system.

  冬天是心脑血管疾病频发的时期，吃一些山楂、黑木耳、苦瓜、玉米、荞麦、西红柿、芹菜等，可以补充维生素和纤维素，对于降血脂也有辅助作用。

  longan, cashew, gorden euyale seed and Chinese yam. In addition, cardiovascular or cerebral vascular diseases occur in winter. Hawthorn, black fungus, balsam pear, corn, buckwheat, tomato and celery with rich vitamins and cellulose help to decrease blood fat.

> 大雪

每年12月6、7或8日，太阳运行到达黄经255°为大雪。古籍《三礼义宗》记载："大雪为节者，行于小雪为大雪。时雪转甚，故以大雪名节。"大雪是相对于小雪而言，到了这个时节，雪往往下得较大，降雪范围也广。《月令七十二候集解》记载："至此而雪盛也。"大雪，顾名思义，雪量增大。

大雪分为三候：一候鹖鴠不鸣，到了大雪时节，寒号鸟都停止了鸣叫；二候虎始交，老虎到了发情的季节，开始求偶；三候荔挺出，一种叫做荔的兰草破土而出，开始生长。大雪时节的降雪量虽然增长，但是全国大部分地区的降水量却减小了，天气较为干燥，这样的气候为农作物越冬创造了极佳的条件。一方面，大雪覆盖地面，使土地温度得

> Major Snow (*Daxue*)

Major Snow begins on the 6th, 7th or 8th of December every year when the Sun arrives at the celestial longitude of 255°. According to *Codex of Three Ritual Traditions*, Major Snow was named because compared to Minor Snow it snows greatly harder in a wider range of places in this season. The *Collection of Lunar Seventy-two Pentads* also describes that "it snows heavily in this season." As the name implies, the Major Snow has the greatest amount of snowfall.

Major Snow has three pentads. In the first pentad, one cannot hear the brown long-tailed pheasant any more. In the second pentad, tigers begin to mate: this is the heat period and people see courtship in tigers. In the third pentad, the *Li* grass breaks through the soil: *Li*（荔）is the ancient name of Chinese Hemp Agrimony. Although the amount of snowfall increases during Major Snow,

• 黑龙江牡丹江雪乡风光
Snow-covered Landscape of Mudan River in Heilongjiang Province

以留存；另一方面，积雪中含有大量的矿物质和微量元素，在天暖融化时灌溉田地，滋养了土地。人们常说"瑞雪兆丰年"，农谚中也有"今年麦子雪里睡，来年枕着馒头睡""冬雪一层面，春雨满囤粮"的说法。立冬时北方的农民进入农闲时节，人们在家中造塘修仓，为来年农事做准备。而南方的农民还

most places in China have less rainfall. This benefits the growth of field crops: firstly, the thick snow cover keeps the temperature of the field warm; secondly, snow particles contain rich minerals and trace elements, which help water and enrich the field. People believe that a timely fall of snow promises a fruitful year. There are also farmers' proverbs going like: this year wheat sleeps under

不能休息，要做好小麦、油菜等农作物的越冬御寒工作。

习俗

在中国南方地区，一到大雪节气，家家户户忙于腌制腊肉，用八角、桂皮、花椒、白糖、盐等熬制成原料涂抹在鸡鸭鱼肉上，然后放到大缸中进行腌制，脱水入味。等到了时日再从缸中取出，用绳线捆了挂在通风的屋檐下晾干。腌肉逐渐发展成为人们迎接新年的一项习俗，甚至还有"小雪腌菜，大雪腌肉"的俗语。

在山东地区，大雪时节有喝红黏粥的习俗。天寒地冻，正值农闲

the snow, the next year the farmer sleeps on steamed bread; and heavy snow in winter brings adequate rainfall in spring. Peasants in the north now have the winter leisure to dig ponds and fix barns in order to prepare for the next year, while their counterparts in the south are still busy with frost prevention work for winter wheat and rape.

Customs

In South China, all families start to make cured meat with star anise, cinnamon, pepper, sugar and salt. Those cured chicken, duck, fish and meat are kept in a jar for dehydration. Then they are taken out of the jar and hung under the eave. Cured meat has become part of the tradition to welcome the New Year as a traditional saying describes: one cures vegetables on the Minor Snow day and meat on the Major Snow day.

In Shandong Province, people have sweet potato porridge during Major Snow. The weather freezing the outside world, people in heavy cotton coats stay at home and make porridge with sweet potatoes and coarse grains.

● 腌菜
Pickled Vegetables

• 《连年有余》年画
New Year Painting: *Rich in the Successive Years*

时节，人们穿上厚厚的棉衣，不再进行频繁的户外活动，在家中将红薯和杂粮熬制成黏糊糊的粥，喝了取暖充饥。

大雪节气处于农闲时期，又距离农历新年很近。这时城镇、农村开始印年画，贴年画成为人们迎接新年、祈求幸福吉祥、盼望丰收的民俗活动。

健康

大雪时节天气寒冷，万物潜藏，人们的日常生活也应顺应自然规律，以"藏"为主。在生活起居上要注意保暖，宜早睡晚起，夜眠

During Major Snow, people rest in winter leisure and prepare for the lunar New Year. In urban and rural areas, people buy and post New Year paintings to express their wishes for happiness and harvest.

Health

As all creatures on Earth hibernate in the cold weather during the Major Snow, one should conform to this natural law, which gives priority to preservation. In terms of one's health, keeping warm and having more sleep (i.e. sleep early and wake up late) are important. Extra bed quilts at night are necessary to keep the warm and blood circulation unobstructed.

时要多加棉被，保持四肢温暖，血液畅通。此时是心脑血管疾病、感冒、支气管炎等病症的高发期，要谨慎起居，适度活动，增强对气候变化的适应能力。在饮食方面，有"冬天进补，开春打虎"的说法。在食物的选择上不要单一地吃肥腻、热量高的食物，应有针对性地调理身体，适当吃一些富含糖分、脂肪、蛋白质和维生素的食物，提高人体的免疫力，补充身体所需的能量，促进新陈代谢、滋养身体。

This is the season when cardiovascular or cerebral vascular diseases, cold and bronchitis intensify, therefore one needs to keep a regular rest time and adequate exercise to improve one's adaption in case of seasonal diseases. In terms of diet, there is a traditional saying: winter supplement gives one the energy to hunt for tigers in spring. Rather than only taking in fatty and high-calorie food, one should also make appropriate adjustments to one's diet based on one's physical condition, such as food rich in sugar, fat, protein and vitamins, which enhances the immune system, refuels energy, promotes metabolism and benefits one's health.

- 穿着冬衣的蒙古族人
  Mongolians in Winter Clothes

> 冬至

冬至是二十四节气中非常重要的节气，早在2500多年前就被人们用土圭观测太阳测定出来，是我国古代人民最早确立的节气之一。每

● 黑龙江牡丹江雪景
Snow-covered Landscape of Mudan River in Heilongjiang Province

> **Winter Solstice (*Dongzhi*)**

Winter Solstice is one of the most important solar terms. It was determined through measuring by ancient Chinese by means of the Sun observation with earthen tablet 2,500 years ago. As the earliest officially recognized solar term, it starts from around 22nd of December every year when the Sun reaches the celestial longitude of 270°. The ancient Chinese interpreting of Winter Solstice is quoted as "only as the *Yin* energy reaches its climax, does the *Yang* energy begin to grow; because the Sun arrives at the southernmost position, the daylight duration is the shortest and the Sun's shadow is the longest". It means that the Winter Solstice day has the shortest day and longest night. After this day, the direct sunlight point moves northward, thus the *Yang* energy recovers and the length of day time grows, which is why

年12月22日左右，太阳运行到达黄经270°为冬至。所谓冬至，古人的解释为："阴极之至，阳气始生，日南至，日短之至，日影长之至。"意思是说，到了冬至这天，白昼最短，黑夜最长。但是冬至之后太阳直射地面的位置又开始向北回归，阳气慢慢回升，白昼时间开始增长。因此又有"冬至一阳生"的说法。

冬至分为三候：一候蚯蚓结，冬至时节天气寒冷，蚯蚓在地下被冻成一团；二候麋角解，麋是一种珍稀的兽类，冬至之后，麋的角到了自然脱落的时候；三候水泉动，地下的泉水开始向上冒热气。民间有"冬至不过不冷"的说法，中国大部分地区1月份都是最冷的月份，天文学上把冬至作为北半球冬季的开始。冬至过后，将进入全年最寒冷的时期，也就是人们常说的"进九"。民间有"冷在三九，热在三伏"的俗语。

习俗

冬至是中华民族的传统节日，称为"冬节""长至节""亚岁"

it is said that Winter Solstice brings the *Yang* energy back to life.

Winter Solstice has three pentads. When the first pentad starts, earthworms curl up: due to the freezing temperature, earthworms roll up under the ground. When the second pentad begins, elk drops their antlers: elk, a type of precious wild animal, naturally drops their antlers after Winter Solstice starts. When the third pentad comes, spring water moves: underground hot spring water steams during this period. There is a folk saying going like "it is only after Winter Solstice that the weather hits the lowest temperature". Most places in China begin their coldest days in January, while the subject of astronomy claims Winter Solstice as the start of winter. After this solar term begins, the northern hemisphere enters into the coldest season, which is called entering the winter nine-day period. People believed that the bitter winter starts from the last *Jiu* (*Jiu* refers to three nine-day periods after the Winter Solstice day, and the last *Jiu* refers to the third nine-day period) and the bitter summer starts from the last *Fu*.

Customs

Winter Solstice, a traditional Chinese

等。周朝时，人们将冬至这天视为岁首，天子要率领文武百官到郊外举行迎岁盛典。到了汉代，人们把冬至作为"冬节"，朝廷要放假"贺冬"，百姓们也要准备食物祭祀祖先，与亲朋宴饮。宋代时，人们对冬至节的重视达到了顶峰，将冬至与春节相提并论，有"冬至大如年"的说法。冬至前后的三天，帝王歇政，民间歇市。到了明清时

festival, is also called the Winter Festival, Long-arriving Festival or Sub-year. In the Zhou Dynasty (1046 B.C.-221 B.C.), the Winter Solstice day was treated as the beginning of the New Year. The emperor would take civil and military officials outside the capital city to host New Year ceremonies. In the Han Dynasty (206 B.C.-220 A.D.), this day was treated as an important festival: the officials had a day off in celebration of

- **北京天坛圜丘坛**

天坛位于北京故宫东南方，是明、清两朝帝王祭天之所。圜丘坛是皇帝举行祭天大礼的地方，坛平面呈圆形，共三层，皆设汉白玉栏板。顶层中心的圆形石板叫做"天心石"，站在上面讲话，声波被栏板反射，会形成显著的回音。

Circular Mound Altar at Temple of Heaven in Beijing

The Temple of Heaven, sitting to the southeast of the Forbidden City in Beijing, was the place to hold heaven worship ceremonies in the Ming and Qing dynasties (1368-1911). The Circular Mound Altar was where the emperor performed his part of the heaven worship ritual. The plane graph of the altar is round and its all three stories have white marble railings. On the top storey there is a round stone plank, called stone of the heaven's center. If one steps onto the plank and speaks to others, one's voice wave hits the railings and forms a returning echo.

期，皇帝与文武百官在冬至日要到郊外举行盛大的祭天大典，称为"冬至郊天"。民间在冬至这天有祭祖的习俗，人们沐浴更衣，准备丰盛的祭品祭拜祖先。过冬节的习俗一直延续至今，在东南沿海地区依然有祭祖活动。冬至这天当地人备齐祭品去祠堂祭祀祖先，祭典之后，还会大摆宴席，款待前来祭祖的亲朋。

在北方有冬至日吃饺子的习俗，民谚有"十月一，冬至到，家家户户吃水饺"。据说冬至吃饺

the arrival of winter, and people prepared food sacrifices for their ancestors and had a big meal with their friends and relatives. In the Song Dynasty (960-1279), people valued the Winter Solstice day as greatly as how they treated the Spring Festival, from which a saying goes "Winter Solstice is as important as the Spring Festival". From one day before the Winter Solstice day to one day after, all political and economic activities ceased. In the Ming and Qing dynasties (1368-1911), on the Winter Solstice day the emperor took the civil and military officials to the countryside and held a grand heaven worship ceremony, which is called the winter heaven worship at *Jiao* (*Jiao* means countryside and here refers to the Circular Mound Altar at Temple of Heaven in Beijing). Common people also arranged their ancestral worship activities on this day. After taking a bath, they put on clean clothes and prepared abundant sacrifice offerings. This tradition is still followed today. In some southeastern and coastal areas in China, people gather at their ancestral hall with prepared sacrifices on the Winter

• 饺子
Dumplings

子的习俗是为了纪念东汉时期的医圣张仲景而流传下来的。张仲景原为医官,河南南阳人,他告老还乡时正值冬至那天大雪纷飞,他看见百姓饥寒交迫,很多人的耳朵都冻伤了,于是叫弟子搭起医棚,将羊肉、辣椒和一些驱寒药材放在锅里煮,捞出后剁碎,再用面皮包成耳朵似的,下锅煮熟,做成"驱寒娇耳汤"给百姓吃。人们服食后,冻伤的耳朵果然被治好了。后来,人们模仿"娇耳"的做法做成"饺子",或称"扁食"。

健康

冬至是一个非常重要的节气。古语有"冬至一阳生",冬至时阴阳二气转化,阴气盛极而衰,阳气开始萌发。因此,除了遵循"秋冬养阴"的原则,还要注意养阳。在生活起居上,古人说"起居有常,养其神也,不妄劳作,养其精也",意思是说要规律作息,合理安排时间,保证充足的睡眠,不要过度劳累,以养足精气神。同时还要注意防寒保暖,进行适度的体育锻炼。在饮食方面,因为冬季是养

Solstice day. After the worship activity, a great banquet is held to treat attending relatives and friends.

In north China, people have dumplings on the Winter Solstice day. A folk saying summarizes: on the 1st day of the tenth lunar month, Winter Solstice begins, and every family cooks dumplings. It is said that this tradition was created in respect of Zhang Zhongjing, who is honored as the Saint of Medicine and lived in the Eastern Han Dynasty (25-220). Zhang Zhongjing, born in Nanyang City in Henan Province, was a medical who official served in the government. When he retired, he decided to go back to his hometown. It was winter time and he found that many people's ears were suffering from cold injuries. He then asked his apprentice to set up a temporary shed, where he cooked some mutton with chillies and cold-dispelling drugs. Then he chopped up these cooked ingredients and wrapped them with flour pieces into ear-like shapes and cooked them with boiled water. He called his soup the Cold-dispelling and Ear-nourishing Soup and cured the patients with it. Later, people followed his way of processing the stuffing and the flour wrappers, and named them dumplings or dumpling soup.

Health

Winter Solstice is a very important solar term from the perspective of the health preservation subject. An ancient Chinese saying is quoted as Winter Solstice brings the *Yang* energy back to life. It means that during Winter Solstice, the *Yin* energy hits the peak and starts to wane while the *Yang* energy begins to recover. As traditional Chinese medical science teaches, nurturing and preserving the *Yin* energy of the human body are important during autumn and winter. However, the *Yang* energy also needs to be nurtured. In terms of daily life, it is suggested that keeping a regular rest time nurtures one's spirit and avoiding overwork nurtures one's energy. At the same time, it is also important to put on adequate clothes and do some sports from time to time. In terms of what to eat, beef, mutton and chicken are good choices to replenish one's vitality and lotus seed, white fungus and date help to bring down inner heat and dissolve excessive grease.

● 火锅（图片提供：图虫创意）
Hotpot

精蓄锐的最佳时机，而补肾、补脾胃是此时的重点。在食物的选择上，牛肉、羊肉、鸡肉等食物能够滋补身体；而莲子、银耳、大枣等食物能够化解火气和油腻。

> 小寒

小寒是表示气温冷暖变化的节气。每年1月6日左右，太阳运行到达黄经285°，为小寒。《月令七十二候集解》中记载："十二月节，月初寒尚小，故云。月半则大矣。"意思是天气已经很冷，但是尚未冷到极点，因此称为"小寒"。小寒之后，中国大部分地区进入了全年最寒冷的时段——"三九"。"三九、四九冰上走""小寒、大寒冻作一团"等民谚，都是形容这一时节的寒冷的。

小寒分为三候：一候雁北乡，古人认为大雁顺阴阳而迁徙，阳气初动，所以大雁从南方开始向北方迁移；二候鹊始巢，北方的喜鹊感觉到阳气回升，开始筑巢；三候雉始雊，野鸡也感受到阳气的升腾开始活跃起来，在山野中行走鸣叫。

> Minor Cold (*Xiaohan*)

Minor Cold reflects the change in temperature among the twenty-four solar terms. This solar term begins around 6th of January every year when the Sun comes to the celestial longitude of 285°. It is described in *Collection of Lunar Seventy-two Pentads* that "the cold is not getting extremely severe until the mid-twelfth lunar month", which explains the reason why this solar term was named Minor Cold. After the Minor Cold day, most places in China entered the coldest days, namely the last three nine-day period after the Winter Solstice day. To describe the bitter cold of this period of time, traditional sayings are quoted as for the third and fourth nine-day periods in winter, we walk on the frozen path; during Minor Cold and Major Cold, we curl up when sleeping.

• 内蒙古阿尔山雾凇
Rime Scenery on Arxan Mountain in Inner Mongolia

Minor Cold has three pentads. In the first pentad, wild geese fly back to the north: ancient Chinese believed that wild geese conformed to the *Yin* and *Yang* energy, thus they fly back to the north where the *Yang* energy begins to recover. In the second pentad, magpies begin to nest, which was also believed to be influenced by restoring the *Yang* energy. In the third pentad, pheasants get active and people hear them chirping in the mountain. As the temperature may affect the growth of crops, timely frost and cold prevention work are vital to protect wheat, fruit trees, vegetables as well as livestock.

Customs

In the past, Chinese people held sacrifice activities to worship the gods and ancestors during Minor Cold, which is called the Winter Sacrifice, or the *La* Sacrifice. *La* refers to the transition from the old to the new, therefore people also call the twelfth lunar month the Month of *La* as the Winter Sacrifice takes place in

由于小寒时节气温很低，对农作物的危害较大，要做好农作物的防寒防冻工作，防止小麦、果树、瓜菜、畜禽等遭受冻寒。

习俗

我国古代有在小寒时节祭祀众神、祖先的习俗，称"腊祭"。腊有新旧交接的含义，腊祭所在的

• 腊八粥
Laba Porridge

十二月也称"腊月"。腊祭的习俗早在先秦时期就已经形成，人们用猎获的野兽祭祀神明和祖先，祈求丰收。

小寒是广东地区较为重视的节气。这一天，广东的百姓特意早早起床，将头天泡好的糯米和香米按照比例混合在一起蒸煮，然后将炒好的腊肠、腊肉、花生米、葱白等辅料拌在饭里，做成好吃的糯米饭来庆祝小寒。在江苏南京地区，有小寒时节吃菜饭的习俗。菜饭也称作"咸饭"，是民间的特色美食，一般随着当地物产的不同，将不同

this month. Early in the Pre-Qin period (c. 2070 B.C.-221 B.C.), people held the Winter Sacrifice with hunted animals to worship the gods and ancestors and prayed for harvest.

People in Guangdong Province value Minor Cold greatly. They get up early on this day to cook soaked glutinous rice and fragrant rice and mix some sausage, preserved meat, peanut and onion stalks into the cooked rice to celebrate Minor Cold. In Nanjing in Jiangsu Province, people have vegetable rice, which is also called salty rice during Minor Cold. The key to cooking vegetable rice is to cook rice along with local ingredients. Vegetable rice in Nanjing is cooked with yellow dwarf chicken, sausage, and salted preserved duck and is served exclusively during the Minor Cold season; while people in Tianjin have Chinese cabbage instead: during Winter Solstice, people place cabbage hearts back to the field and cover them with manure. Those ripened cabbages won't be ready to eat until Minor Cold, when all stored vegetables are in short supply.

菜肴与主食一起烹制。南京菜饭是用南京特产的矮脚黄、香肠、板鸭与糯米饭一起烹煮而成的，味道鲜美香甜，是小寒时节特有的食物。天津地区在小寒节气有吃黄芽菜的习俗。黄芽菜是白菜的一种。冬至时节，百姓们将白菜心留下来继续栽种，并用粪肥覆盖以保持一定的温度。等到小寒时节，蔬菜匮乏，黄芽菜却正好可以吃了。

健康

小寒时节已经进入三九时期，是一年当中最冷的季节。中医认为寒易伤人体阳气，为防御寒冷气候对人体的侵袭，在生活起居上，要注意保暖，预防感冒。在阳光普照的时候，要适当进行户外锻炼，强身健体，增强御寒能力。在饮食方面，中医讲究"养肾御寒"，可以多食用一些温热食物来补益身体。

Health

The last nine days of Minor Cold come with the lowest temperature of the year. According to the theory of traditional Chinese medicine, cold affects the *Yang* energy in the human body, therefore keeping warm is important. Adequate exercise on sunny days also helps enhance one's health against the cold. With regards to the daily diet, traditional Chinese medical science suggests nurturing one's kidney against the cold, for instance, warm food.

• 雪中乡村
Village in Snow

> 大寒

　　每年的1月20日或21日，太阳运行到达黄经300°为大寒。大寒是相较于小寒而言的，是一年中最寒冷的时期。古籍《授时通考》引《三

> **Major Cold (*Dahan*)**

Major Cold starts on 20th or 21st of January every year when the Sun reaches the celestial longitude of 300°. Compared to Minor Cold, Major Cold

• 黑龙江伊春冬韵
Winter in Yichun City, Heilongjiang Province

礼义宗》对大寒的解释为："大寒为中者,上形于小寒,故谓之大……寒气之逆极,故谓大寒。"

大寒分为三候:一候鸡始乳,到了大寒时节,阳气慢慢充盈,母鸡开始孵化小鸡;二候征鸟厉疾,正是鹰隼等猛禽捕食能力最强的时候,它们在空中盘旋寻觅猎物,以补充能量抵抗严寒;三候水泽腹坚,水面的冰被冻得结结实实的,尤其是水域中部的冰最坚硬、厚实。大寒时节天气寒冷,干燥多风,降雪积而不化,北方大地一片冰天雪地的严寒景象。但由于冬季即将结束,天气虽然依旧寒冷,南方已经隐隐能够感受到大地回春的迹象。大寒节气里各地的农活较少,北方地区农民积肥为春耕做准备,同时加强牲畜的防寒防冻工作。南方地区则加强小麦及其他作物的田间管理。

习俗

民间有"过了大寒,又是一年"的说法。这里的"年"指的是农历新年,因此大寒时节的民俗活动都充满了浓浓的年味。人们奔

is getting much colder. In fact, it is the coldest period of a year. The book *A General Study on Issue: Almanac of Farming Activities* has quoted the explanation of Major Cold from *Codex of Three Ritual Traditions*: Major Cold, starting from the mid-twelfth lunar month, is a more freezing solar term compared to Minor Cold just as its name suggests. People named this solar term the Major Cold due to its extremely cold air.

Greater Cold has three pentads. In the first pentad, hens start laying eggs. In the second pentad, predator birds circle in the sky: they hunt for food all day to get enough calories to survive in winter. In the third pentad, the groundwater freezes into solid ice, especially in the center of lakes and the middle stream of rivers. During Major Cold, the weather is freezing, dry and windy, and snow does not melt. However, as winter is close to its end, spring is just around the corner in southern China. Fieldwork in this season is relatively less heavy: people in the north collect manure that will be used during spring ploughing; while people in the south intensify the field management of winter wheat and other crops.

• 杨柳青木版年画《新年多吉庆》
Yangliuqing Wood-block New Year Painting: *Happy New Year*

波劳碌了一年，到了年末岁尾，开始为过年做准备。打扫房屋，清扫庭院，购置新衣裳、鞭炮、春联等年货，制作腊肠、腊肉、干粮等食物，祭祀祖先、举办宴席。每年的腊月二十三为小年，在这一天有祭灶的习俗。在民间传说中，灶王神是玉皇大帝封的"九天东厨司命灶王府君"，专门负责管理各家的灶火。每年岁末，他就会回到天宫向玉皇大帝禀报工作。百姓为了让他

Customs

A traditional saying goes as "The (lunar) New Year comes when Major Cold ends". Therefore festival activities taking place in Major Cold all share a sense of New Year celebration. People have worked for a whole year and now it is the time to celebrate their achievements for themselves: cleaning houses and yards, New Year shopping for new clothes, firecrackers and Spring Festival couplets, cooking sausage, preserved meat and

春节

　　春节又称"阴历年",俗称"过年",是中国最隆重热闹的传统节日。春节起源殷商时期岁末年初举办的祭祀神明、先祖的活动,并作为一个传统节日流传下来。春节有辞旧迎新的特殊意义,在历史发展中形成了许多流传至今的风俗习惯,从腊月初八就开始举行腊祭,熬制腊八粥,腌制腊八蒜。腊月二十三祭灶之后,采办鸡鸭鱼肉、瓜果蔬菜、新衣新帽等年货,还要沐浴更衣、打扫卫生。腊月三十为除夕,各地在腊月三十下午有祭祖的风俗,并贴门联等,还要挂上红灯笼。除夕之夜,人们燃放烟花爆竹,准备丰盛的菜肴,全家欢聚一堂吃年夜饭,守岁迎新。北方的年夜饭主食为饺子,南方吃汤圆。还有许多有象征意义的菜肴,如鱼,取年年有余之意。到了大年初一,人们走亲访友,相互拜访祝福。在新年期间,各地还举办庙会,开展舞狮、舞龙、扭秧歌、杂耍等各种民俗活动。

- 年画《大过新年》中的包饺子
  Dumpling Making in Chinese New Year Painting Having a Good New Year Gathering

● 张灯结彩的地坛庙会
Temple Fair at Temple of Earth

Spring Festival

The Spring Festival, also called the Lunar New Year or more commonly, the New Year, is the most important and popular traditional festival in China. Its creation dates back to the year-end god and ancestral worship activities early in the Shang Dynasty (1600 B.C.-1046 B.C.) and now it has become a cultural tradition. The Spring Festival, now considered a ritual to welcome the New Year's coming, is celebrated with a series of activities: on the 8th day of the twelfth lunar month, people hold Winter Sacrifice and cook Laba porridge and Laba garlic; after the God of Kitchen worship activities, people carry out New Year shopping for food and clothes, taking a bath, putting on clean clothes and cleaning their house; on the New Year's Eve day (30th day of the twelfth lunar month), people have ancestral worship ceremonies, post spring couplets and hang lanterns; in the evening, people get together with their family members to have a big meal and set off strings of firecrackers. In the north, people eat dumplings at the New Year's Eve dinner while in the south people have dumpling balls. Some ingredients have been imbued with special meanings suggesting good luck or happiness, for instance, fish signifies being brimful of happiness; from the 1st day of the first lunar month, people visit their relatives and friends with gifts and good wishes; and during the New Year Festival, there are temple fairs open to the public where people enjoy lion and dragon parades, yangko dance and variety shows.

年画《灶王爷》
New Year Painting: God of Kitchen

other food, worshipping ancestors and holding banquets. On the 23rd day of the twelfth lunar month, people celebrate Preliminary Eve and offer sacrifice to the God of Kitchen. In Chinese folk legendary stories, the God of Kitchen was appointed by the Emperor of Heaven in charge of all Chinese kitchens. At the end of the year, he returned to the heavenly palace and reported to the emperor. To make the God of Kitchen speak well in front of the Emperor of Heaven, people offer incenses and sacrifices including sugar melons made of malt syrup and flour in front of his enshrined portrait on the 23rd day of the twelfth lunar month.

Health

Major Cold is the coldest solar term in winter when all creatures hibernate. There is a traditional folk saying that during the Major Cold, one drinks ginseng and Astragalus liquor in the morning and eats lyceum rehmannia pills in the evening against the cold air. It is in this season that cardiovascular or cerebral vascular diseases and respiratory diseases frequently happen. Therefore one with weak health should better stay indoors and keeping warm is important when

在玉皇大帝面前说好话，每年腊月二十三会向设在灶壁神龛中的灶王爷画像敬香，并供上用饴糖和面做成的糖瓜等供品。

健康

大寒时节正处于冬季最为寒冷的阶段，万物蛰伏。民间有"大寒大寒，防风御寒；早喝人参黄芪酒，晚服杞菊地黄丸"的说法。心脑血管疾病和呼吸系统疾病在这种

*《月曼清游图之围炉博古》陈枚（清）

Chinese Painting: *Evening Tour: Conversation Next to the Stove*, by Chen Mei (Qing Dynasty, 1616-1911)

严寒而干燥的天气下容易复发，因此在生活起居方面，身体虚弱的人最好尽量待在温暖的室内，如果一定要出门也要注意保暖御寒。要早睡晚起，心境平和，以调养自己的神气。在饮食方面，大寒节气是进补的时节，动物脂肪与药膳相结合能起到较好的效果。多吃紫苏叶、生姜、大葱、辣椒、花椒、桂皮等发散风寒的食物，对于抵御寒冷也能起到很好的作用。而此时生冷、燥热的食物与气候相悖，因此还是少食为妙。

doing outdoor activities. To maintain and promote better health, one should have adequate sleep and a peaceful mind. In terms of what to eat, Major Cold is a good time to have nourishing food, especially animal fat cooked in a medicated way. Cold-dispelling food including purple Perilla leaf, ginger, spring onion, pepper, Chinese prickly ash, and cinnamon is effective in warming one's body. Moreover, one should take in less cold and raw food or inner-heat boosting food during Major Cold.